Piezoelectric pyramid
The clean Hydrogen power of ancient Egypt
Terry Lee

Terry Lee

Contents

Introduction

Acknowledgements

Together we stand on the shoulders of the giants who came before us, too many to mention here but a special Thank you to the genius of Nikola Tesla. His vision and pioneering work gave rise to the modern world and continues to point the way to our future developments. Much gratitude for our popular media platforms and the virtual buffet of modern content creators who present these concepts of science and physics with entertaining and easy to understand presentations.

This text presents an expose' of the Great Pyramid's true purpose and function. The many illustrations and easy to use scientific references cast a focused light into the lost technology of our distant past and its strategy of operation by exploiting the properties of water, steam and HHO fuel. Power generation via the piezoelectric effect coupled with the high voltage ionization properties of Hydrogen gas, along with evidence of wireless receivers and user devices from the old world.

Also a special thank you to our prominent authors and researchers for sharing their knowledge and ideas about so many subjects. Their efforts and hard work continue to resonate and inspire the next generation of researchers like ourselves.

Hypothesis: A proposed explanation made on the basis of limited evidence as a starting point for further investigation.

Theorem: A general proposition not self evident but proved by a chain of reasoning. A truth established by means of accepted truths.

The internal combustion, piezoelectric, electrochemical hypothesis of the Great Pyramid of Giza

From a layman's perspective in the study of this building, one can't help but notice that the Queen's chamber seems designed to hold water and the Queen's chamber Niche just has a look, like it contained something hot just above the water line. Having said that, the hypothesis and theory of operation arises from the appearance of wear and tear evidence present on key features of the Great Pyramid, practical reasoning and the implications of physics.

The Great Pyramid is a piezoelectric power generator, Hydrogen / Oxygen fueled, internal combustion. The Queen's chamber held water and liberated fuel through pyrolysis and electrolysis, contained a heat source in the Niche and the "air shafts" are high voltage conductors. The King's chamber is made of piezoelectric material and the sarcophagus contained the igniter.

Come along for the ride as we explore this hidden technology of the past through illustrations, scientific references and a few antique photos that reveal the true purpose and function of the Great Pyramid, also known as Cheops or the pyramid of Khufu.

Synopsis

This text presents an argument for the presence of an advanced technological civilization from our distant past. Perhaps when we see the pyramids, we are looking at some of the few remains from the cataclysm upon the earth known as the *Younger Dryas event* about 12,000 years ago.[1]

Basic physics and insights from a recently declassified book written in 1963, [2, 3, 4] suggests that when the polar regions become heavy with ice, the spinning ball of the earth (or oblique spheroid if one prefers) can roll onto it's side or even more terrifying, perhaps the earth's crust can slip against the molten mantle and the poles move to the equator in the span of just a few hours. All of civilization destroyed in a single day from the mile high tsunamis and supersonic winds that race across the lands causing catastrophic destruction and upheaval around the planet. This occurrence would also explain the rapid rise in sea levels and drop in global temperatures that researchers have been unable to explain with existing theories about the *Younger Dryas event.*

The theory of operation for the Great Pyramid is presented along with the energy values and volume equivalences for water, heat and steam. The electrolysis energy of water and production of Hydrogen / Oxygen gas, the use of plasma HHO production and a practical set of values for dynamic operation of the system. An effort is made to use plain language and basic concepts of physics, common knowledge or at least commonly available knowledge in the twenty first century.

So please grab a favorite beverage and return your seat to a reclined position while we begin and visit the author's web page @ terryleeauthor.carrd.co to see the dynamic operation in action with a full motion GIF.

[1] www.wikipedia.com "Younger Dryas"

[2] Chan Thomas, author "The Adam and Eve story" 1963 Bengal Tiger Press

[3] www.cia.gov "Chan Thomas, Adam and Eve story" declassified and sanitized 2013

[4] www.youtube.com The Why Files 2023 "CIA classified book about the pole shift, mass extinction and the true Adam and Eve story"

Giza plateau

Chapter 1
Old world, high tech

From the truly massive stones of Baalbek, Lebanon to the polygonal construction at Cusco and Machu Picchu, the oldest stonework at the bottom of the stack displays the greatest capabilities and technical prowess of the ancient builders.

From the remarkably precise stonework preserved in the Quorikancha to recent laser scan studies of predynastic vase work recovered from the pyramid of Djoser, it is the oldest works that display the highest levels of precision in the hardest materials.[1]

The Great Pyramid is only one example of advanced high energy technologies present in our distant past, albeit not the most readily apparent. Perhaps the most visible, yet hidden in plain sight example of high tech from the past would be the ancient remains of what appears to be some version of resonant cavity device the size of a house, located atop the hill at Sacsayhuaman, just outside the city of Cusco, Peru.

The ancient builders of Cusco, Sacsayhuaman, Ollantaytambo and Machu Picchu clearly utilized an advanced stone working technology to great effect. An understanding of *this* device should tell us what we want to know about how the ancients were able to manipulate stone like it was putty. A few examples are included here for reference, but that's a subject for another time as this book will focus on the operation of the Great Pyramid.

[1] www.youtube.com Uncharted X 2023 "Scanning a predynastic vase to 1,000[th] of an inch.

Cusco, Peru

Sacsayhuaman

Hilltop at Sacsayhuaman

Great Pyramid, internal cavities

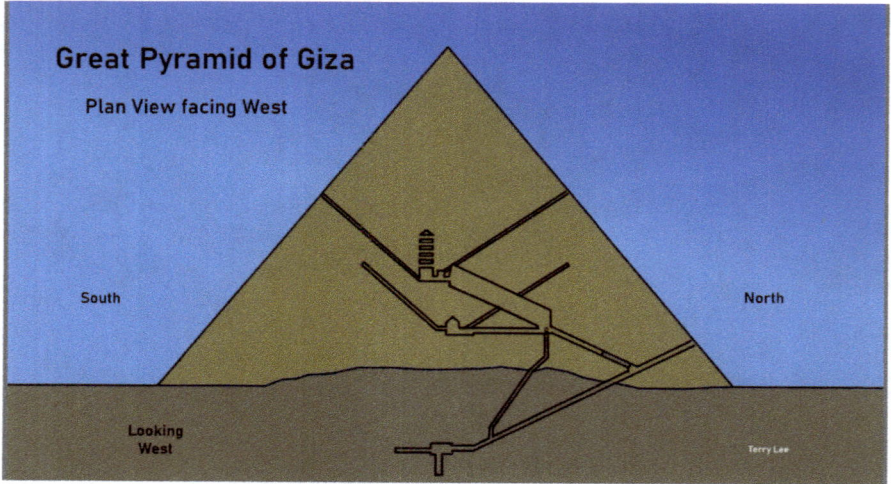

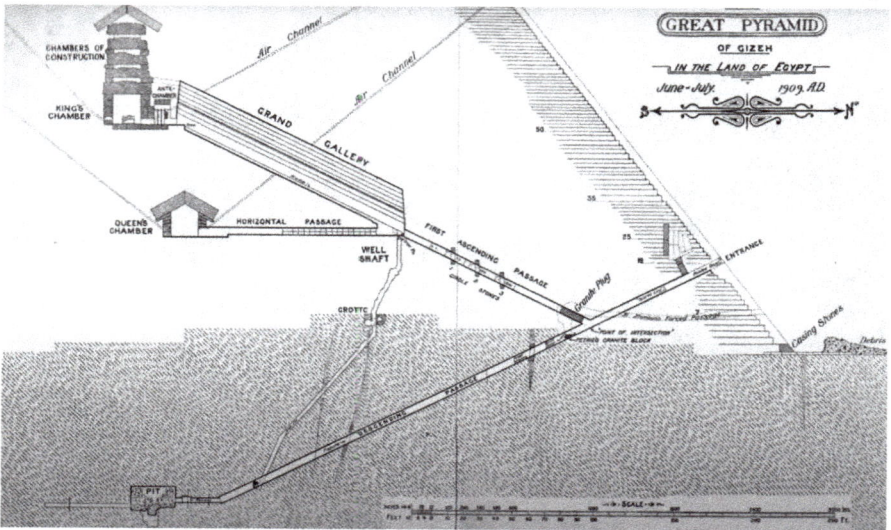

Great step, circa 1910

System components

1 Steam chamber, water to fuel converter

1a Heat source, steam powered electrostatic generator

2 Water reservoir

3 Drain

4 Combustion chamber

4a Igniter

5 Red granite, piezoelectric blocks

6 High voltage conductor tubes

7 Electrodes

8 Scoop

9 Signal beacon

9a Iron plate

10 Vent

11 Limestone insulator

12 Gas capacitance

13 Conductive final

14 Gas cavity amplifier, diffuser

15 Tuning chamber

Component identifier

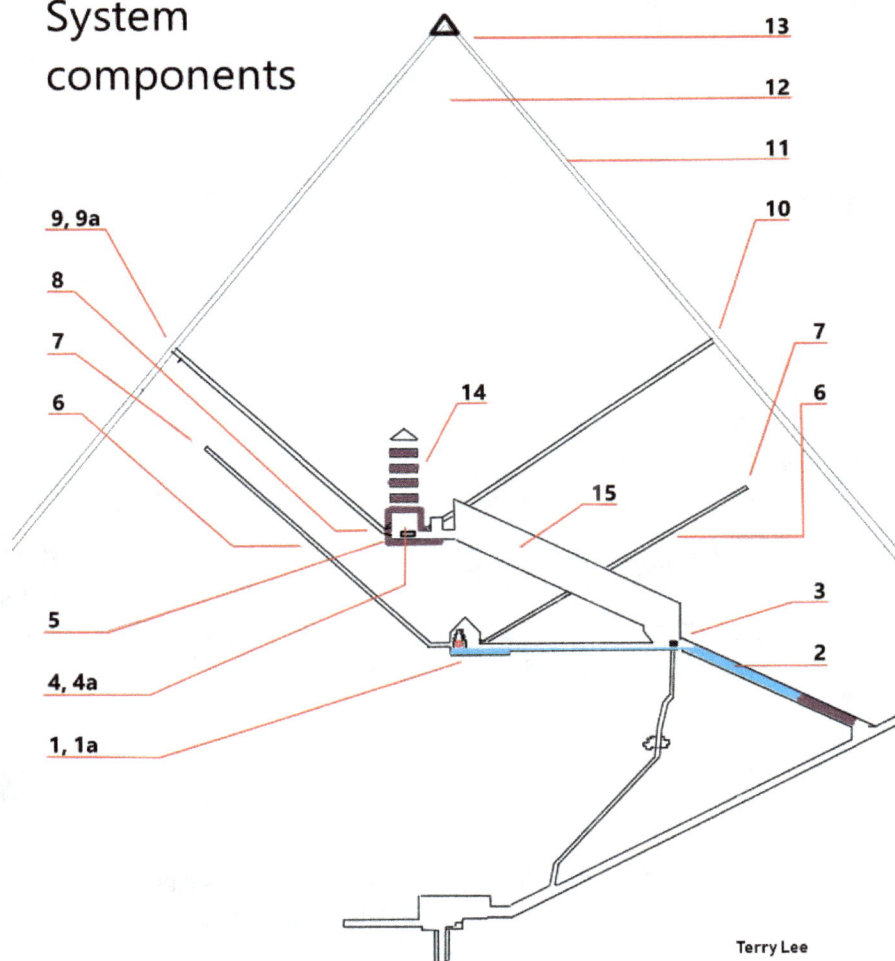

System components

13
12
11
10
9, 9a
8
7
6
7
6
14
15
5
3
4, 4a
2
1, 1a

Terry Lee

Chapter 2

The engine of the Great Pyramid

Tuning chamber

Since the system primarily operates as an internal combustion power plant, it seems reasonable to begin with the Grand Gallery and its function as a tuning chamber. Somewhat similar to the purpose of a *"Stinger"* type exhaust system equipped on high performance two cycle dirt bike engines.

The shape of the Stinger exhaust is designed to reflect the pressure wave of the exhaust pulse back to the exhaust port.[1, 2] The length of the Stinger exhaust is calibrated so that the reflected pressure wave arrives back at the exhaust port at the desired moment in time, corresponding to the peak power design speed of the engine. Similarly, the length of the Grand Gallery provides a clue to the system's frequency of operation, which works out to about 4 hertz, or 240 cycles per minute. See system operation on page 16.

[1] www.youtube.com Steps to Podium 2020 "How 2-stroke exhaust pipes work"

[2] www.youtube.com Driving 4 answers 2024 "How two stroke exhaust pipes really work"

Two cycle Stinger exhaust pipe

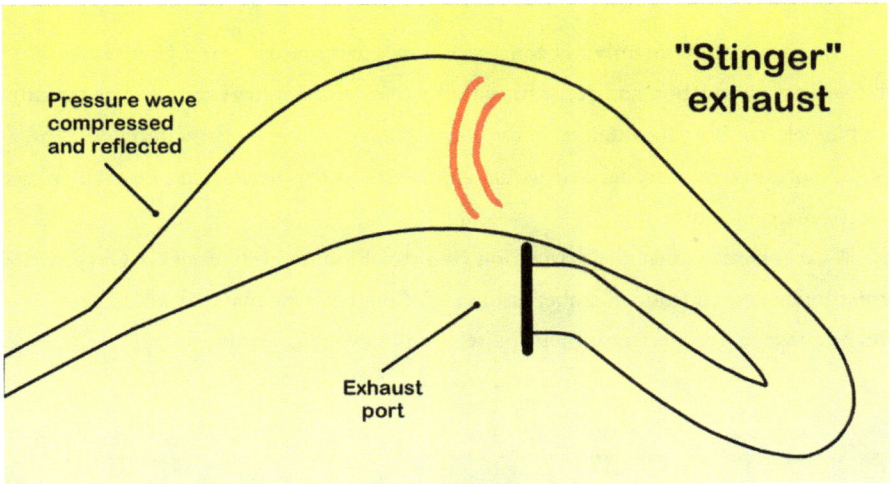

Pressure wave compressed and reflected

"Stinger" exhaust

Exhaust port

Grand Gallery tuning chamber

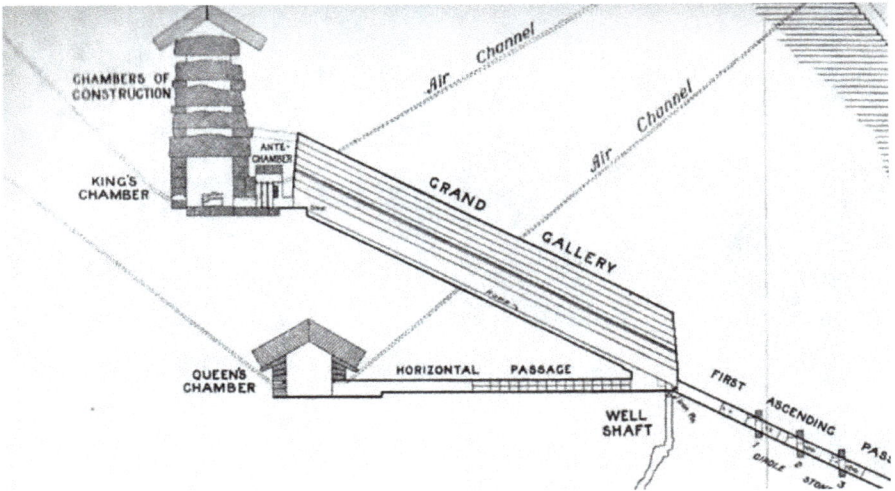

CHAMBERS OF CONSTRUCTION

Air Channel

Air Channel

ANTE-CHAMBER

KING'S CHAMBER

GRAND

GALLERY

QUEEN'S CHAMBER

HORIZONTAL   PASSAGE

FIRST

ASCENDING

WELL SHAFT

PAS

Combustion chamber

The combustion chamber is constructed of quartz bearing, red granite blocks of rather large proportion and constitutes the system's main voltage source. The quantity of piezoelectric crystal contained in the combustion chamber construction is unclear, as is the portion of the crystal that would contribute to the generation of a high voltage electromagnetic pulse.

What *is* clear, is that the detonation of a stoichiometric Hydrogen / Oxygen fuel mixture in a combustion chamber fashioned of piezoelectric material would generate a high voltage and an electromagnetic pulse of unknown magnitude.

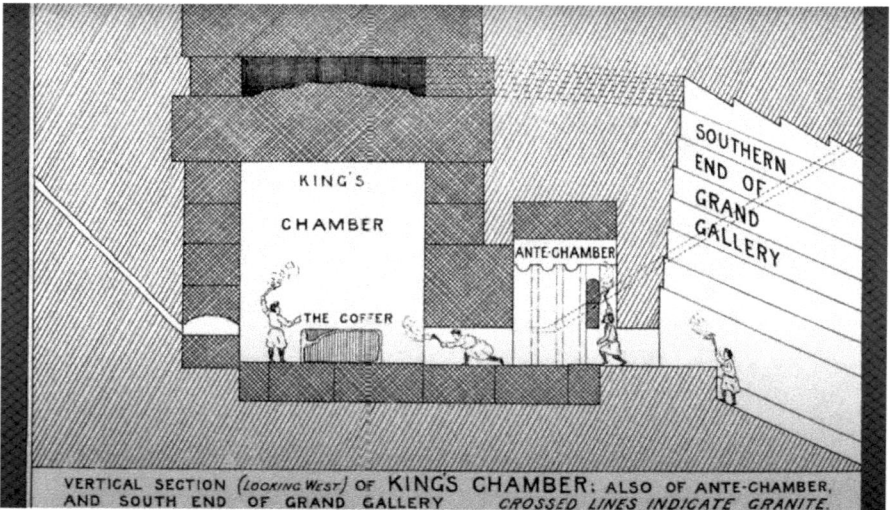

VERTICAL SECTION *(LOOKING WEST)* OF KINGS CHAMBER; ALSO OF ANTE-CHAMBER, AND SOUTH END OF GRAND GALLERY   *CROSSED LINES INDICATE GRANITE.*

The igniter

The igniter containment is located near the northwest corner of the combustion chamber, this location becomes significant when considering the propagation of the flame front and pressure wave within the combustion chamber. Since the nucleation point of the detonation occurs at the northwest corner of the chamber, the focal point would be located at the opposite southeast corner where the south wall shows extensive damage around the southern shaft.

The peoples choice award

The igniter containment box is of course empty, leaving much room for speculation about what exactly was there. Here the hypothesis suggests that the igniter consisted of a capacitor containing a radioisotope, beta voltaic current source such as Radium 226 with an energetic decay tree and half life of 1600 years, providing a sufficient rate of charge through the electron emission of Beta Decay.[1,2]   The rate of charge, capacitance of the capacitor and the distance of the spark gap calibrating the frequency of the spark output and the system's operation.

System operation

The length of the Grand Gallery provides a clue to the system's design speed, or frequency of operation. Curiously enough, the speed of sound [3] increases with higher temperature and humidity levels to about 440 meters per second at 100% humidity and 100°C. It's worth mentioning that the humidity level of the Grand Gallery would likely exceed 100% to some degree of supersaturation due to the continuous and voluminous production of steam from the steam chamber.

Each combustion event in the combustion chamber would produce a pressure pulse that travels the length of the Grand Gallery and then reflects back to its source. Since the distance between the combustion chamber and the bottom of the gallery is just under 60 meters, the pressure pulse would take about 250 milliseconds to make the round trip, suggesting an operating frequency of about 4 hertz. It would seem optimal for this time frame to coincide with the time it takes for combustion chamber pressure to drop to a relative vacuum after each combustion event due to rapid cooling of the steam, thus helping to fill and swirl the combustion chamber for the next combustion cycle.

[1] www.wikipedia.com "Beta decay"

[2] www.energy.gov US Dept. of Energy "DOE explains Beta decay"

[3] www.wikipedia.com "Speed of sound"

A water vapor environment

It presents a twist on physics to consider the humidity level or water content of the air, since all of the air would be purged from the building and only steam, water vapor and HHO fuel gases would occupy the internal spaces of the building. The thermal capacitance of water vapor along with the temperature / pressure relationship with water and the behavior of steam would feature prominently in the operation of the combustion chamber.

When the HHO fuel gas reacts and burns to form water in the combustion chamber, a shockwave is produced due to HHO having the highest specific impulse of all known fuels, with a reaction speed of 3,000 meters per second. The reaction is exothermic and the newly formed water molecules along with some of the surrounding water vapor would flash to steam, producing a sharp rise in combustion chamber pressure. The water vapor would serve to dilute the fuel gas mixture and act as a thermal capacitance to cool the newly formed steam of HHO combustion. Chamber pressure would then drop to a relative vacuum due to rapid cooling of the steam, thus drawing in a fresh gulp of water vapor and fuel gas mixture for the next combustion cycle.

The space at the top of the Grand Gallery provides a reservoir for the HHO fuel gas and is thought to work in tandem with the Antechamber to accomplish fuel metering.

Chapter 3
The voltage source

Piezoelectric crystal

The quartz crystal contained within the red granite blocks of the combustion chamber serves as the system's main voltage source via the piezoelectric effect.[1, 2, 3]

The Silicon atoms carry a small positive charge while the Oxygen atoms are slightly negative. The symmetric Silicon / Oxygen lattice of the quartz crystal is electrically neutral until a shockwave or vibration causes a distortion of the crystal lattice. When the Silicon / Oxygen lattice is compressed or distorted on the appropriate axis, the crystal becomes more positive at one end and more negative on the other end. This momentary polarization causes a potential difference across the crystal and a flow of electrical current from the negative side to the positive side.

Quartz crystal structure

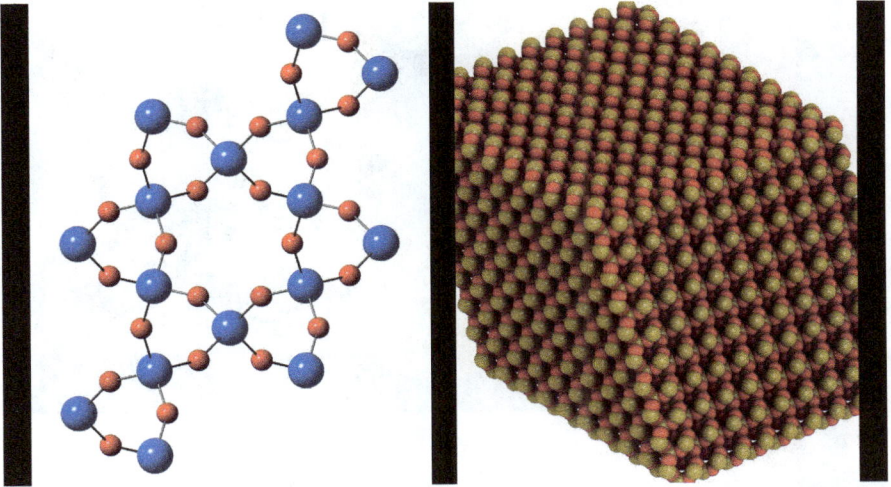

[1] www.wikipedia.com "Piezoelectricity"

[2] www.youtube.com Physics High 2017 "piezoelectric effect explained"

[3] www.youtube.com Steve Mould 2019 "Why hitting crystals makes electricity"

Silicon / Oxygen lattice

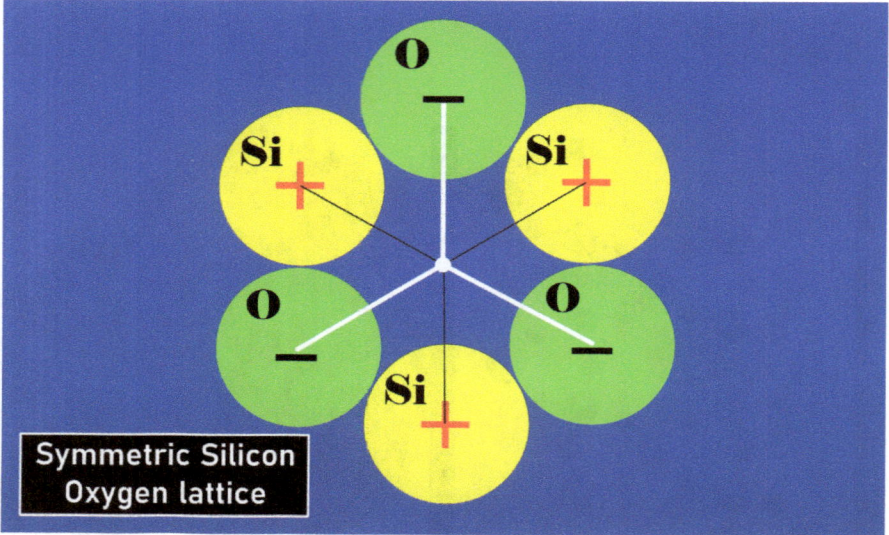

Symmetric Silicon Oxygen lattice

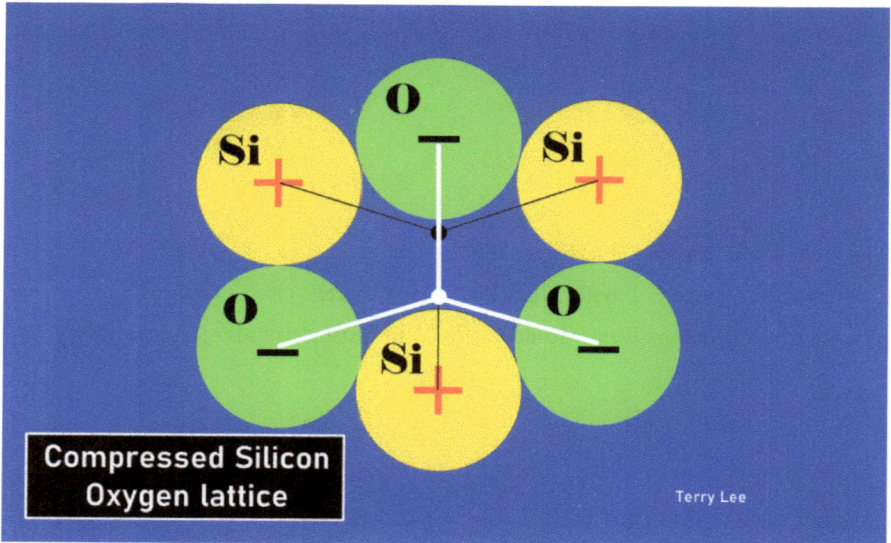

Compressed Silicon Oxygen lattice

Terry Lee

Piezoelectric current

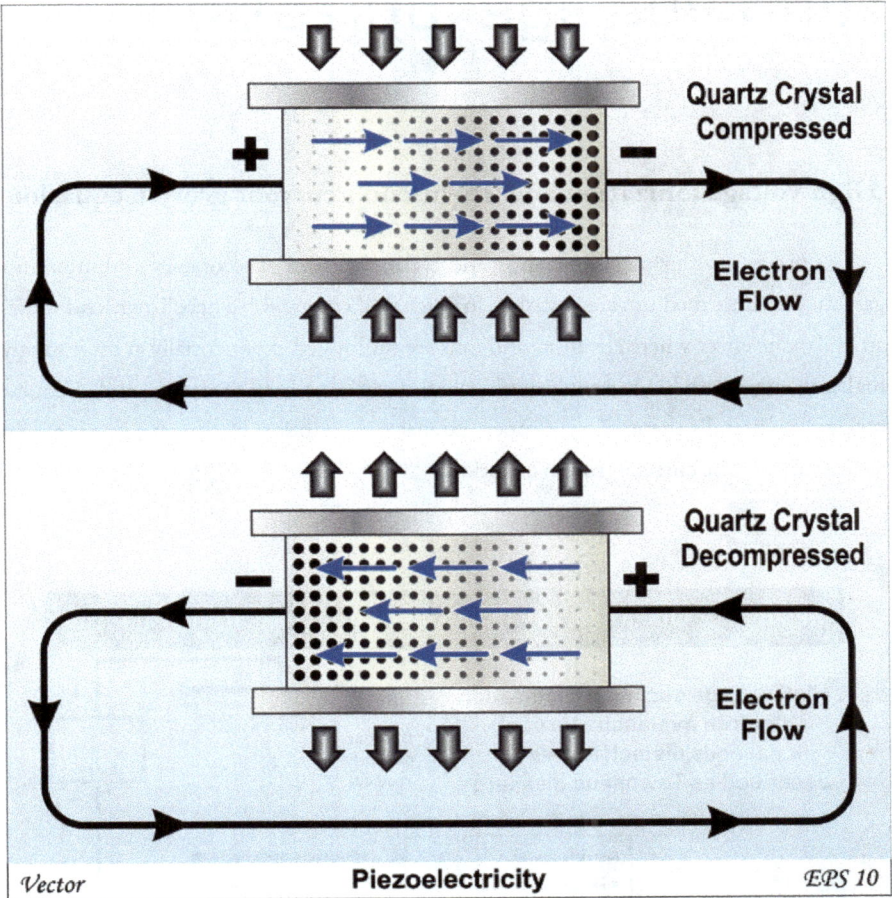

Quartz Crystal Compressed

Electron Flow

Quartz Crystal Decompressed

Electron Flow

Vector   **Piezoelectricity**   *EPS 10*

Chapter 4
High voltage ionization

High voltage ionization and Townsend's current growth equation

The processes of primary ionization and avalanche effect of secondary ionization in a gaseous dielectric medium are described in *Townsend's theorem.*[1] There Townsend shows an avalanche effect where electrons and ions are multiplied exponentially at increasingly higher voltages with his current growth equation.[2,3] Interested readers should of course review Townsend's theory for a proper, extensive description of primary vs secondary ionization and the current growth equation.

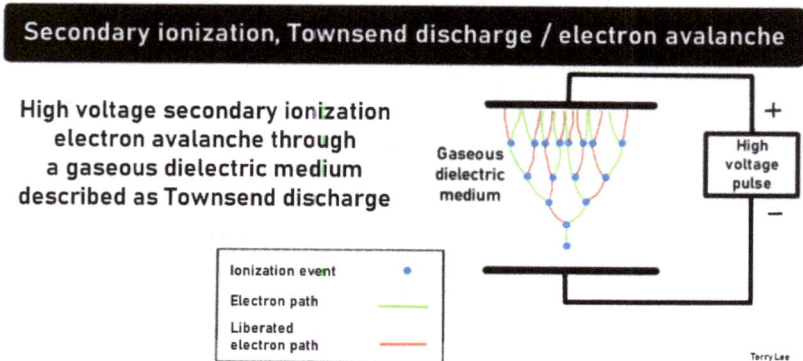

Secondary ionization, Townsend discharge / electron avalanche

High voltage secondary ionization electron avalanche through a gaseous dielectric medium described as Townsend discharge

Gaseous dielectric medium

+

High voltage pulse

−

Ionization event	●
Electron path	
Liberated electron path	

Terry Lee

Ionization in Hydrogen gas[4]

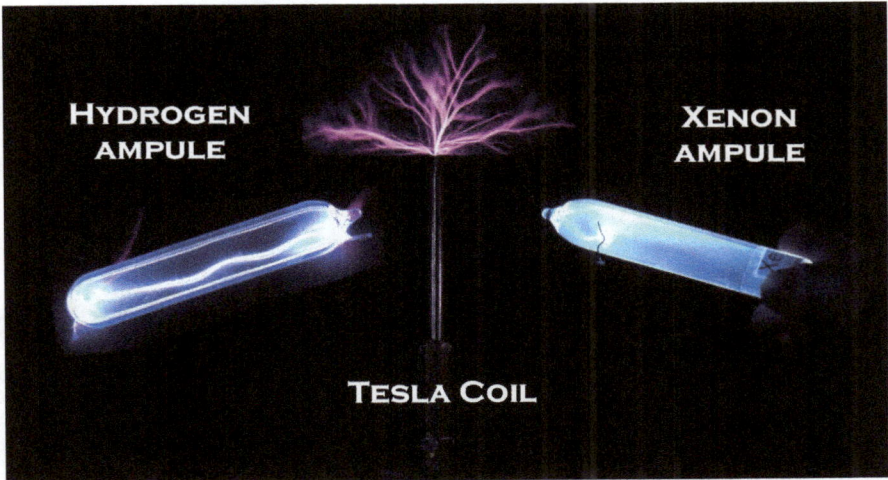

[1] www.youtube.com Sai Academy of engineers 2021 "Townsend's primary and secondary ionization coefficient"

[2] www.wikipedia.com "Current growth through secondary ionization"

[3] www.youtube.com Sai Academy of engineers 2021 "Townsend's current growth derivation"

[4] www.youtube.com Thoisoi 2 2023 "Hydrogen, the lightest gas in the universe"

Chapter 5
The steam chamber

Steam chamber, water to fuel converter

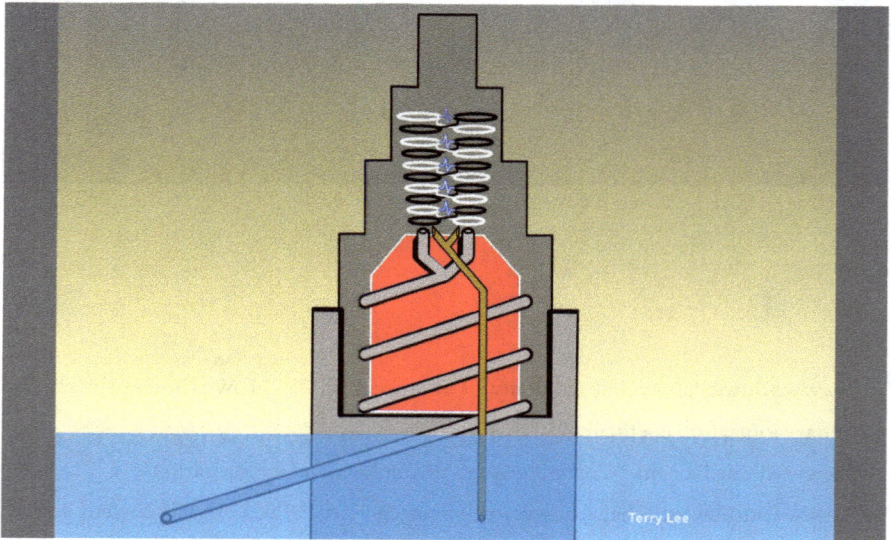

The hypothesis holds that the energy for the system was supplied from a heat source located in the Niche of the Queen's chamber. For the purpose of presenting the theory of operation, it seems prudent to describe the heat source as being compact and efficient. Something akin to the SNAP reactor program developed for NASA in the 1960's to power their early satellites.[1, 2] In addition, the space provided within the Niche would accommodate the hot end of the SNAP reactor unit just about right.

The original heat source is of course missing and while its operation is unknown, the hypothesis suggests that the heat source drew a stream of water from the pool and heated the water to a superheated steam. The steam was mixed with liquid water via a venturi tube and directed through a series of coil sets. This arrangement would constitute a steam powered electrostatic generator and serve to produce high voltage electrical arcs through the steam causing some of the water molecules to decompose through pyrolysis [3] and electrolysis,[4] producing Hydrogen and Oxygen gasses. The parallel concept in physics is demonstrated by the famous experiment known as *Kelvin's water dropper experiment* or *Kelvin's thunderstorm generator.*[5, 6]

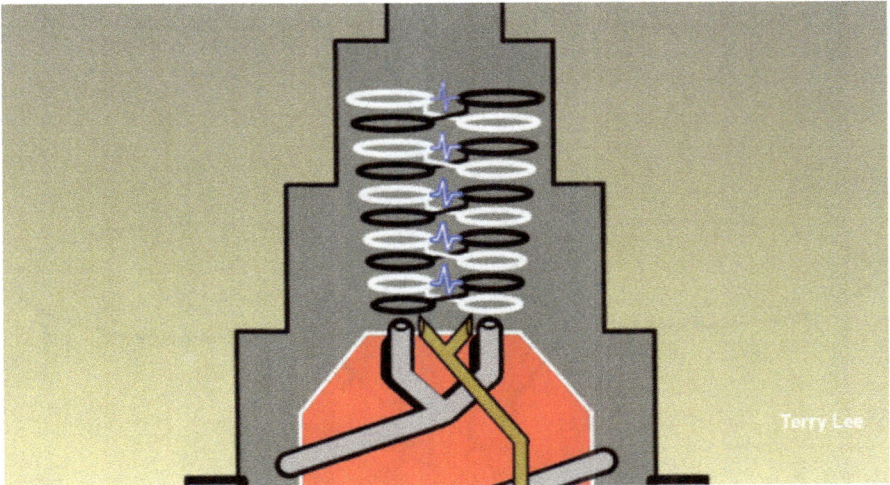

[1] www.wikipedia.com "Systems for nuclear auxiliary power"

[2] www.youtube.com Periscope film 2015 "First nuclear reactor in space, SNAP 10a program 1965"

[3] www.wikipedia.com "Pyrolysis"

[4] www.wikipedia.com "Electrolysis"

[5] www.wikipedia.com "Kelvin's Water Dropper"

[6] www.youtube.com Veritasium 2014 "Sparks from falling water, Kelvin's thunderstorm"

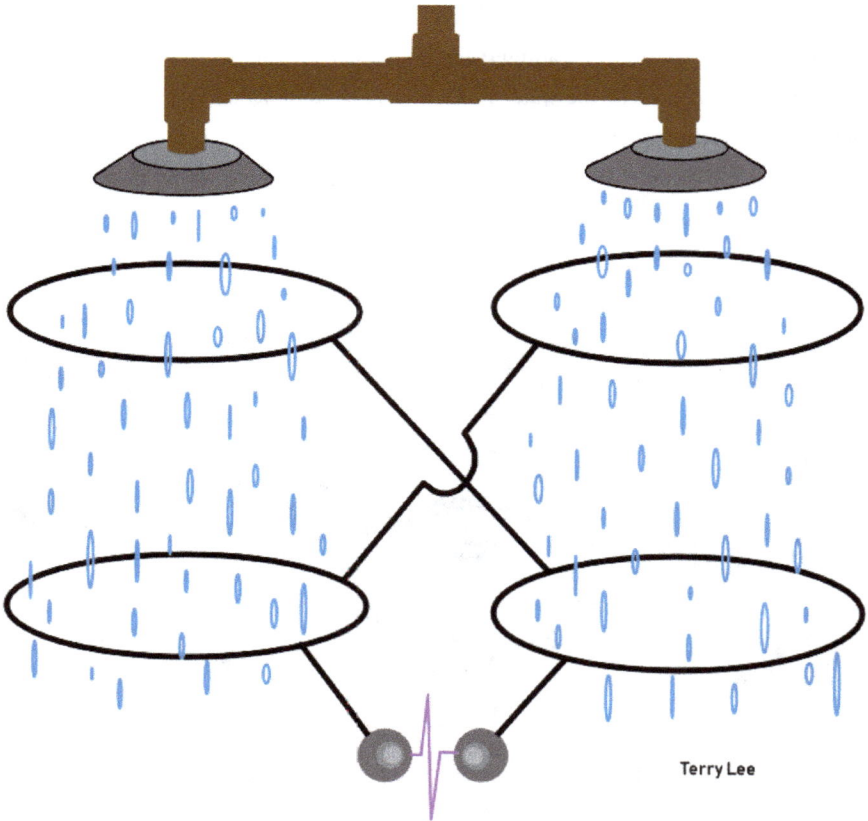

Terry Lee

Kelvin's Water Dropper experiment

Also known as Kelvin's Rainstorm, or Lord Kelvin's Thunderstorm Generator.
The experiment shows that dribbling water droplets through a set of cross
connected coils will produce high voltage electrical arcs of 10kv – 20kv
between the electrodes.

Queen's chamber, 1904

QUEEN'S CHAMBER, 1904

High voltage tubes

The high voltage conductor tubes of the steam chamber angle upwards through the building to the north and to the south of the chamber from the building's centerline. Both of the tubes terminate at a stone plate featuring a pair of metallic looking projections with the color of corroded Copper. The last few feet of the tubes are constructed from a different type of stonework than the rest of the tube, suspected to be the same low Magnesium Tura limestone used for the building's exterior casing.

The upward angle and cavity space of the tubes present another place for the light-weight H^2 gas to be displaced into while it leaks through and permeates the stonework through every available crack or crevice on its path upward. The tubes become conductive when the high voltage EMP [1] sweeps past and ionizes the gaseous dielectric medium contained within the tubes and upwardly surrounding stonework.

During a combustion event, a high voltage electromagnetic pulse is generated and propagates from the voltage source in all directions. The hypothesis suggests that the voltage is sufficient to produce the avalanche effect of secondary ionization, the EMP carries electrons and grows in strength as it sweeps through the gas content of the building like a wave.

The voltage source is offset from the centerline of the building to the south side and the southern shaft is somewhat closer to the voltage source than the northern shaft. As the wave of electrons propagates through the building it reaches the southern shaft before it reaches the northern shaft.

---

[1] www.wikipedia.com "Electromagnetic pulse (EMP) Also referred to as a transient electromagnetic disturbance"

Top end of Queen's chamber Northern shaft

Gantenbrink's door, Northern shaft [1]

Top end of Queen's chamber Southern shaft

Gantenbrink's door, Southern shaft [1]

[1] www.youtube.com Ancient Architects 2021 "Exclusive: First look inside the Queen's chamber northern shaft"

When the high voltage EMP sweeps through the building, the avalanche effect of secondary ionization strips electrons which are carried along with the wave per *Townsend's current growth equation*, leaving positively charged ions in its wake. This causes the southern shaft to become positively charged at the same moment the wave of electrons reaches the northern shaft, creating a large potential difference between the shafts and across the steam chamber.

The hypothesis suggests that the potential difference is sufficient for the abundance of electrons to jump the 5.7 meter gap across the chamber to the positively charged south side. The resulting electrical arc through the steam within the chamber producing HHO fuel through pyrolysis and electrolysis, perhaps best described as *Plasma HHO production*.

The resonant property [1] of the King's chamber also provides a clue to the frequency of electrical arcs produced through the steam chamber. Given the dimensions of the combustion chamber and the speed of sound at 100% humidity and 100°C, the acoustic frequency of the chamber works out to 72 hertz north and south, while the east west component runs at 36 hertz. The velocity of the *P wave* through granite is roughly 6 kilometers per second, which suggests a 3 kilohertz vibration frequency for the walls of the chamber at an estimated one meter thick.

While the intensity of the resonance would quickly die out with each back and forth of the oscillation, this combination suggests an intermittent to nearly continuous production of electrical arcs through the steam chamber, somewhere in the neighborhood of 3,000 cycles per second.

---

[1] www.youtube.com Ancient Presence 2017 "Epic acoustics in the Great Pyramid"

Steam chamber, water to fuel converter

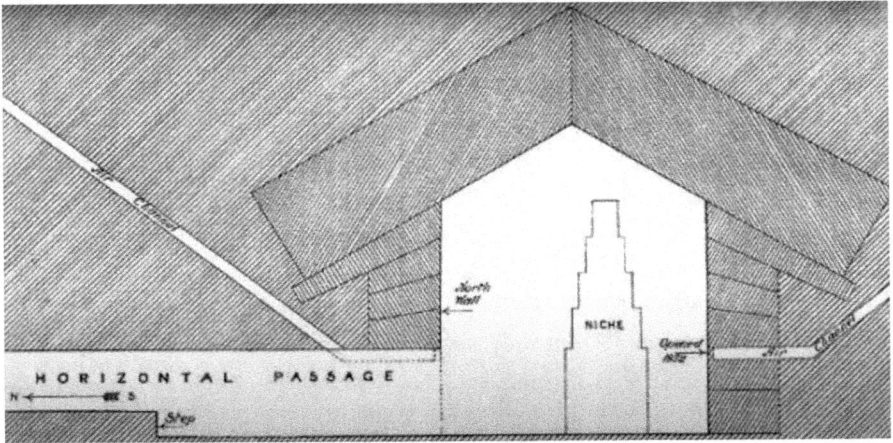

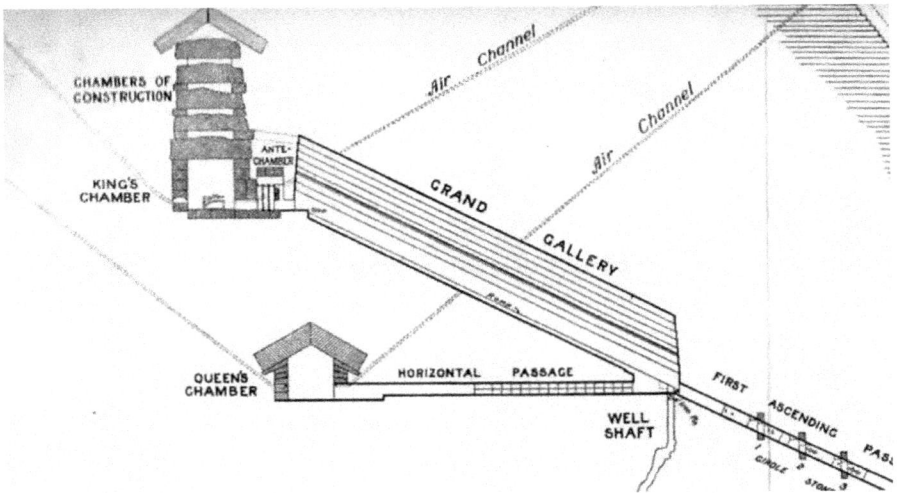

Chapter 6
Volumes and values

Water, plus heat and electricity

While quantifying the energy relationships of the system seems daunting at best, the things we know about physics provide a solid framework for an understanding of the Great Pyramid's operation. For example, we know that it takes 2.3 kilowatts of heat to drive the phase change of one liter of water from a liquid to a gas, producing 1.65 cubic meters of steam. We also know that it takes 13.2 megajoules, or 3.67 kilowatts of electrical energy for one hour to drive the electrolysis of one liter of water to produce 1.2 cubic meters of Hydrogen gas and 0.6 cubic meters of Oxygen.

We know the pyrolysis temperature of water to be about 3,000*C, or 5,400*F and the temperature of high voltage electrical arcs can far exceed this decomposition temperature of water.

Quantity of heat, volume of steam

Dynamic operation of the steam chamber's electrostatic generator calls for a robust heat source and substantial volume of steam production to serve as the system's energy source and primary driver. While the original values are long lost over the vast sea of time, the things we know about physics again provide a solid framework to establish a reasonable set of values for dynamic operation of the system's primary driver.

Starting with the basic unit of one liter of water plus 2.3 kilowatts of heat to produce 1.65 cubic meters of steam. Plus something more than a SWAG (sort of wild ass guess), but an educated guess about the energy level needed to drive the system's operation. The baseline presented here suggests that the system would require roughly the amount of heat needed to boil 4 liters, or about one gallon of water per second to produce 6.6 cubic meters of steam per second. This rate of steam production is enough to create a 5 meters per second velocity of steam through the steam chamber's horizontal passage, enough of a production rate to fill the roughly 2000 cubic meter interior volume of the building in about five minutes. The basic unit of heat needed to drive the phase change of one liter of water is 2.3 kilowatts of heat, suggesting a minimum output for the heat source of about 10 kilowatts once the system warms up to full operating temperature.

The SNAP 10a reactor from the 1960's presents a compact heat source and real world comparison with an initial heat output of 30 kilowatts, suggesting that this category of device and amount of heat would be practical to power the system's primary driver.

One cubic meter of water heated to a boil will produce 1,650 cubic meters of steam.

Volumes

The interior dimensions of the cavity spaces within the structure are well known, complex and boring to present so only the volume of the interior cavities is presented here and estimated to be somewhere between 2000 and 2200 cubic meters of space above the level of the drain.

Below the level of the drain, the horizontal passage leading into the Queen's chamber steps down a full half meter. The depth of the step combined with the surface area of the floor, gives the Queen's chamber a liquid water holding capacity of about 16 cubic meters or 16,000 liters of water. Opposite the Queen's chamber, the water reservoir has an estimated capacity of about 65 cubic meters or 65,000 liters of water, giving the system a total capacity of about 81,000 liters of liquid water. The drain plumbing and drain cavity are thought to act as a buffer or silencer during operation and serve as a drain path for excess water, then excess Oxygen during the system's initialization and gas purge phase.

Electrolysis, plasma and Hydrogen / Oxygen gas

While we know the energy value for the electrolysis of one liter of water to be 13.2 megajoules and the corresponding volumes of HHO gas to be 1.2 cubic meters of Hydrogen and 0.6 cubic meters of Oxygen per liter of water. Now comes the novel idea of a steam powered electrostatic generator and the unknown efficiency of *plasma HHO production*. This concept seems completely absent from our current paradigm, yet poses a tantalizing possibility for its dynamic potential and invites a new field of research for its promise of sustainable, clean burning fuel from water.

One cubic meter of water decomposed to HHO gas will produce 1,200 cubic meters of Hydrogen and 600 cubic meters of Oxygen.

Practical value set

This practical example suggests that the system utilized enough heat to boil 4 liters of water per second to produce 6.6 cubic meters of steam per second, enough to fill the interior volume of the building in about five minutes. Of course the steam will cool and condense to water vapor, saturating all of the interior surfaces as it condenses to liquid water and recycles back to the steam chamber. While an accurate estimate is difficult to establish due to the varying density of steam condensing to water vapor, a best guess suggests that about 4 cubic meters of water would be in circulation at any one time during system operation.

The volume of Hydrogen gas needed to saturate the stonework is also difficult to estimate, but thought to be about the same as the proper interior volume of the structure. Best guess suggests that about 4 cubic meters of water converted to HHO gas would be needed to saturate the stonework and prime the building for operation at system start up.

This gives a total for the amount of water needed for start up at about 8 cubic meters of water, or about half of the Queen's chamber liquid water capacity.

Once the system starts up and begins running, each combustion event in the combustion chamber would send a pressure wave that travels the length of the Grand Gallery and reflects off the surface of the water in the reservoir. The impact of the pressure wave would cause water to splash out of the reservoir and into the horizontal passage leading to the Queen's chamber, thus keeping the steam chamber filled with water during system operation.

Gas purge

Once the construction and assembly of the system was complete, it seems reasonable that the start up process would begin by filling the system with water and turning on the heat source. Gas production would need to run for some period of time to achieve a partial Hydrogen saturation of the building and enable system start up. Once gas production was running and the air was being purged from the building, the lightweight Hydrogen gas would tend to pool at the highest points and fill the spaces from the top down while permeating the stonework through every available crack or crevice on its path upward.

The Oxygen gas is heavier and would tend to pool at the lowest point where the drain is located. The drain provides a path for the heavier Oxygen to escape through the drain cavity and ultimately out the front door.

Given these conditions, the building would eventually become partially saturated with Hydrogen, the air purged out and the internal cavities filled with steam, water vapor and Hydrogen / Oxygen gasses.

This purging of all the air is of critical importance to avoid the formation of Nitric acid and oxides of Nitrogen due to the presence of high voltage electrical arcs in the steam chamber and the high temperature of combustion in the combustion chamber.

The other interior volume

Equally important and difficult to estimate is the Hydrogen gas capacitance of the building's stonework, particularly in the upper portion of the building and around the Queen's chamber "air shafts". The cracks and gaps between the stones where the H^2 gas can leak and be displaced into while always moving upward in the process.

The remains of the low Magnesium, fine Tura limestone casing suggests that the insulation layer was fit precisely to create a gas tight seal, causing the H^2 gas to be retained as it rises to the highest points available within the structure. This Hydrogen gas capacitance presents a key aspect for the system's operation by containing the gaseous dielectric medium to serve as an electrical conductor at high voltages through primary ionization and act as an amplifier at increasingly higher voltages through the avalanche effect of secondary ionization as described in Townsend's theorem.

Hydrogen energy value references [1, 2, 3]

[1] www.youtube.com Professor M does science 2023 "The Hydrogen atom ground state"

[2] www.youtube.com Professor M does science 2023 "The Hydrogen atom energy spectrum"

[3] www.youtube.com Physics made easy 2016 "Deriving energy levels of the Hydrogen atom"

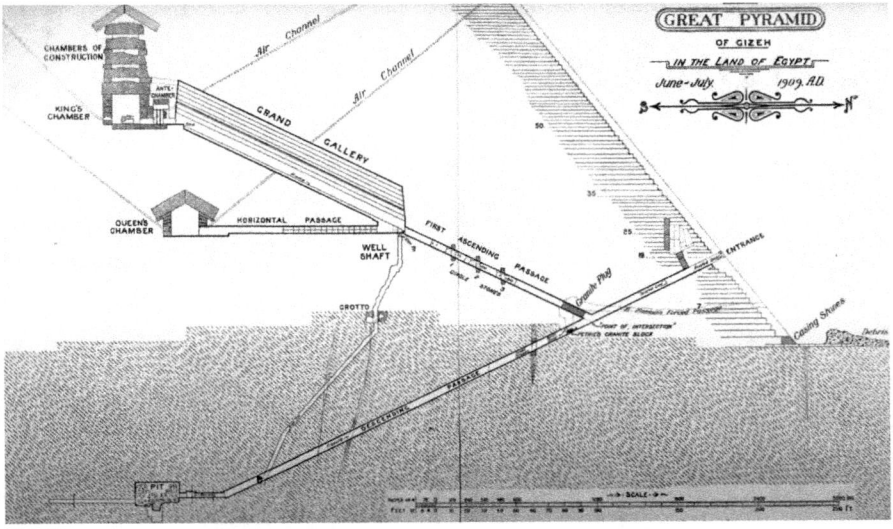

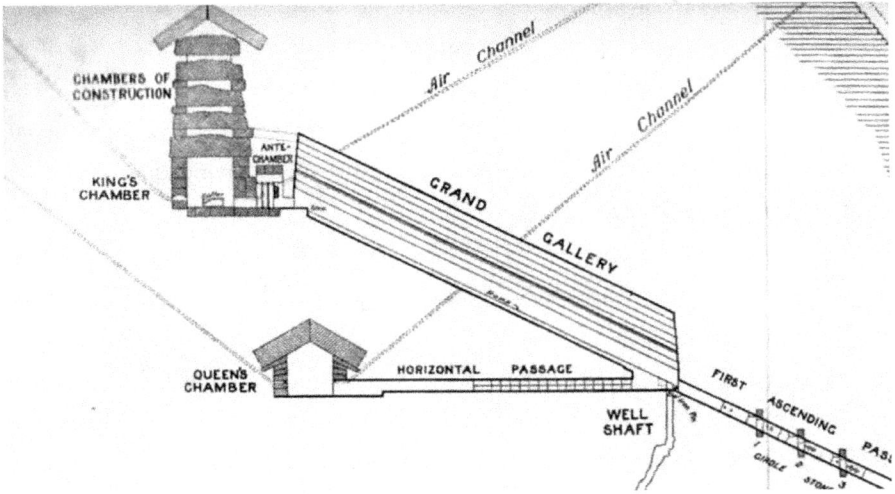

Chapter 7
Power output

Power output and the relieving chambers

The series of cavities and complex of shaped beams known as the "Relieving chambers" would appear to serve two distinct functions for the system's operation. It's of particular interest that each one of the beams in the complex is made flat on the bottom and sides, yet each beam features a unique profile on their top surface. Taken together as a whole functioning component, the apparently random undulations of the beams would serve to scatter or diffuse energy moving vertically through the complex.

More importantly, these chambers present a set of upward cavities for the H^2 gas to be displaced into and collected. This places a series of gas filled cavities directly between the voltage source and the power output.

The theory suggests that when an electromagnetic pulse propagates from the voltage source in all directions, the voltage is sufficient to produce the avalanche effect of secondary ionization through the gaseous dielectric medium.[1, 2] As the EMP propagates through the building, it carries a wave of electrons that grows in strength. The upwardly moving portion of the wave is amplified via the gas filled cavities and funneled to the pyramidion where it yields from the pointy spot as an electrical arc of unknown magnitude.

Gas cavity amplifier

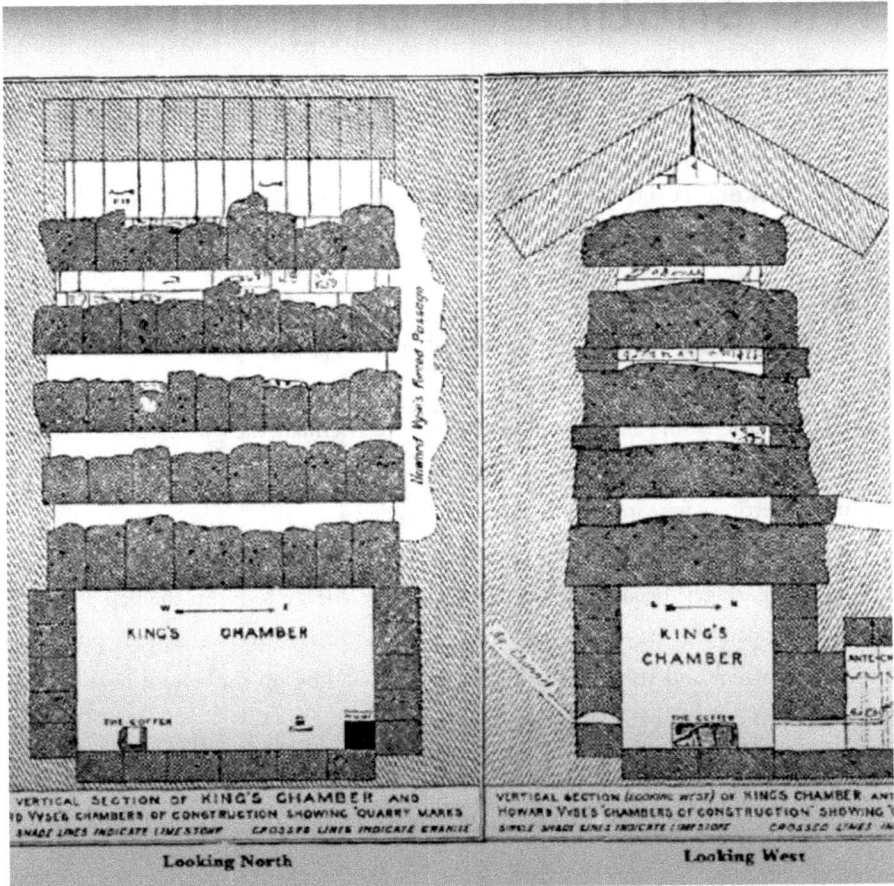

VERTICAL SECTION OF KING'S CHAMBER AND
D VYSES CHAMBERS OF CONSTRUCTION SHOWING 'QUARRY MARKS
SHADE LINES INDICATE LIMESTONE   CROSSED LINES INDICATE GRANITE

Looking North

VERTICAL SECTION (LOOKING WEST) OF KINGS CHAMBER AND
HOWARD VYSES CHAMBERS OF CONSTRUCTION SHOWING
SINGLE SHADE LINES INDICATE LIMESTONE   CROSSED LINES I

Looking West

[1] www.wikipedia.com "Current growth through secondary ionization"

[2] www.youtube.com Sai Academy of engineers 2021 "Townsend's current growth derivation"

Chapter 8
The shoulders of giants

Nikola Tesla, the Tesla coil and the Wardenclyffe tower

We gather from history that Nikola Tesla wanted to operate his Wardenclyffe tower at 8 hertz, apparently he was convinced that his wireless power transmitter would achieve resonance at 8 hertz due to the speed of light and the circumference of the earth. He believed that this would allow people everywhere to attenuate the wireless power with an antenna he described simply as a metal plate. He also said that one could operate their lamp by inserting a wire into the ground, that is all.

In the classic Tesla coil design, a capacitor is charged to a high voltage until the space across the spark gap becomes ionized allowing the charge to jump the gap and flow through the pancake shaped primary coil. This produces an electromagnetic pulse that induces a current and voltage into the secondary winding up the central tower to the toroid shaped capacitor atop the tower.

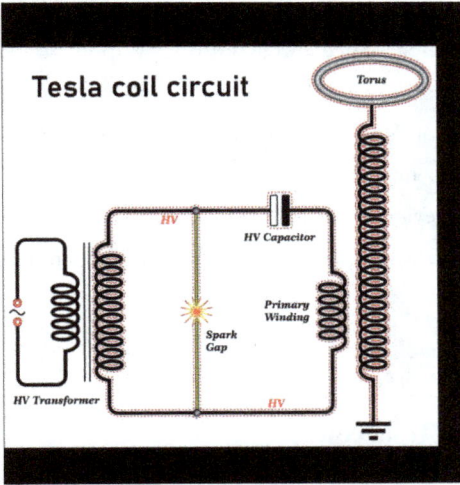

Wardenclyffe tower

The capacitor is toroid shaped to add capacitance and avoid pointy spots where charges will concentrate causing electrical arcs to yield from the pointy spot of the capacitor. Both the primary side and the secondary side are tuned, resonant inductor / capacitor circuits that oscillate at a high frequency.

Interested readers are recommended to review the excellent presentation suggested below for a thorough description and explanation of the spark gap Tesla coil's operation in this six part, do it yourself series.[1]

Spark gap Tesla coil

---

[1] www.youtube.com Diode Gone Wild 2020 "How to build a spark gap Tesla coil"

Credit to our prominent authors and researchers

Much gratitude to our modern community of authors and researchers, their hard work and engaging presentations have shed new light into the hidden mysteries of our forgotten past. Emerging theories and new research are revealing evidence of an advanced technological civilization that flourished in our distant past, the old world before the Great Flood.

New evidence of advanced chemical processing and ammonia production on the Giza plateau.[1]

New research into the energy infrastructure of the ancients and insights into the state of their nuclear power program.[2]

Popular works of literature and perhaps the most prominent theory of recent decades describes the Great Pyramid as an advanced ancient power plant.[3] It's of particular interest where the author carefully documents the dimensions of the King's chamber's southern shaft and its funnel shaped opening. The theory culminates with the suggestion that the southern shaft constitutes a microwave guide and serves as the system's power output.

[1] www.youtube.com The Land of Chem 2025 Episodes of advanced chemical engineering by the ancients

[2] www.youtube.com Ancient Nukes 2025 Episodes explore the ancient energy infrastructure and insights into their nuclear power program.

[3] Christopher Dunn author "Giza Power Plant" 1998 Bear & company

Radiation and the x-ray tube vs the resonant cavity magnetron

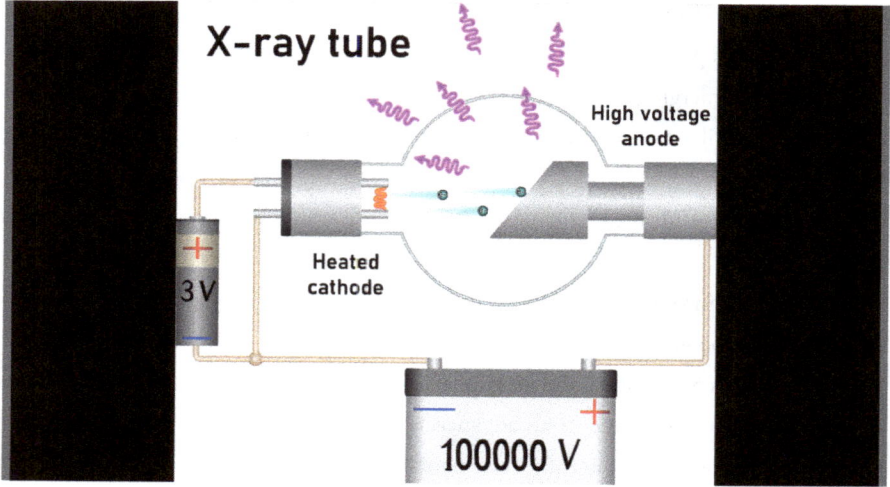

The physics of these two devices provide a relevant example of how radiation would be produced in the King's chamber's southern shaft.

In the x-ray tube, electrons are liberated from the heated cathode due to thermionic emission.[1] The negatively charged electrons accelerate as they are drawn toward the high positive voltage anode and radiation is produced when the high velocity electrons collide with the high voltage anode.

Sort of the opposite is true for the resonant cavity magnetron where the heated cathode is driven to a high negative voltage. The electrons are spun by a magnetic field and accelerate as they are drawn toward the relatively positive, resonant cavity anode which is maintained at a ground potential.[2]

[1] www.wikipedia.com "X-ray tube"
[2] www.wikipedia.com "Cavity Magnetron"

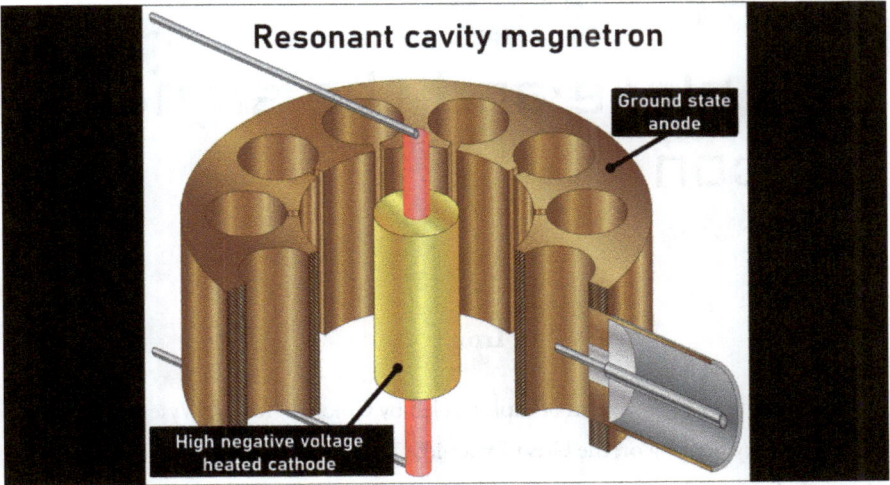

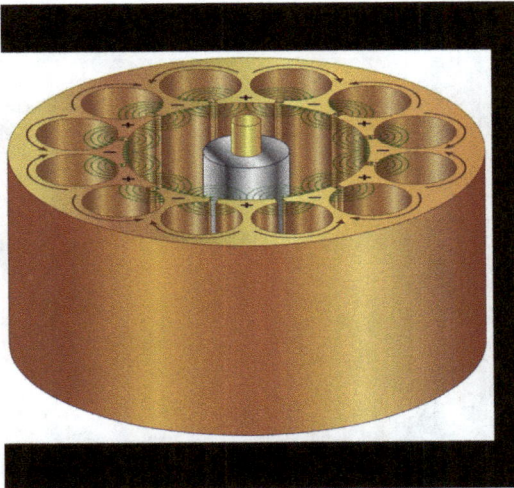

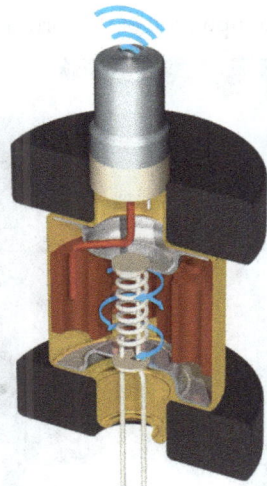

Chapter 9

Iron plate and the signal beacon

Iron plate

In the spring of 1837, a team of explorers led by Colonel Howard Vyse were conducting exploratory blasting on the Great Pyramid. On May 26, 1837 the remains of an Iron plate was discovered by J. R. Hill after blasting the outer two tiers of stone from around the mouth of the King's chamber southern shaft.[1] The remains of the plate were studied and found to be wrought Iron and were also determined to be of ancient origin due to the composition of the corrosion on its surface.

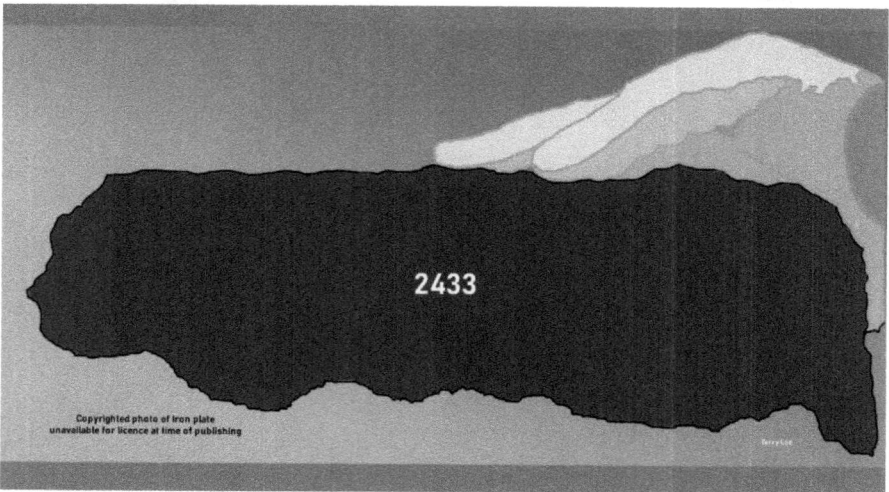

2433

Copyrighted photo of Iron plate
unavailable for licence at time of publishing

Terrykid

[1] www.youtube.com LINES IN SAND 2023 "Investigating the Iron plate found in the Great Pyramid"

Signal beacon

During a combustion event, the hypothesis suggests that some of the unburned fuel gas is forced into and accelerated up the combustion chamber's southern shaft. The shockwave induces a high voltage electromagnetic pulse that ionizes the Hydrogen gas while the expansion of steam causes a sharp rise in combustion chamber pressure. The ionized gas is propelled and accelerated up the shaft, producing long wave electromagnetic energy and a blip of shorter wavelength radiation when some of the high velocity electrons collide with the ground state anode Iron plate located at the top of the shaft.

The Hydrogen ions (Hydrogen atoms that have been stripped of their electron) are positively charged bare protons since the Hydrogen atom does not contain a neutron, except in the case of Deuterium or Tritium and would be attracted to the grounded Iron plate.

The absorption of positively charged protons by the Iron plate would cause a momentary bias, or availability of charge carriers causing the plate to become weakly attractive to the negatively charged, high velocity electrons.

King's chamber south wall

Image courtesy of
Ancient Presence

Chapter 10

Wireless receivers, user devices and the wear & tear evidence

Wireless receivers and user devices

While we're left with plenty of room to speculate about the purpose and function of the massive obelisks from the ancient world, the prime example we have of a wireless receiver and user device are the "Dendera Lamp" relief carvings in the Temple of Hathor.

The artwork is of course an artistic impression that's open to interpretation by anyone who sees it. Here the hypothesis suggests that a portion of the artwork depicts a man holding an ampule of ionizing gas near a wireless receiver, the tall structure with a stack of four disks atop the column known as a "Djed Pillar", commonly seen in Egyptian artwork. It's thought that the receiver attenuates the wireless power and yields a high voltage.[1]

It's well known that holding an ampule of ionizing gas near a high voltage source such as a Tesla Coil will cause the gas within the ampule to glow.[2]

[1] www.youtube.com Ancient Nukes 2025 "Technologies of ancient Egypt part -2- Coils"

[2] www.youtube.com Thoisoi 2 2023 "Hydrogen, the lightest gas in the universe"

Gas ampules

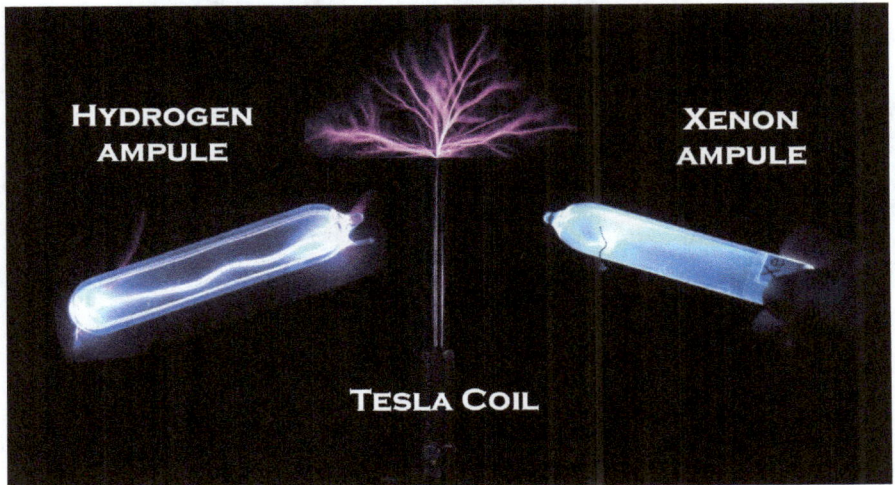

In the artwork, the ampule that the man is holding appears to feature a cable connected to a box. The Hypothesis suggests that the box contains a capacitor and that during operation, a high voltage and small current is exchanged back and forth between the receiver and capacitor downstream of the ampule. The small current moving back and forth through the ampule at four hertz causes an electrical arc to propagate and extinguish four times a second, leading to the artistic rendering of a wiggly snake within the ampule.

Temple of Hathor

Djed pillar

Wear & tear evidence

The limestone Great Step

The substantial erosion and erosion patterns present on the limestone "Great Step" is of particular interest since it shows two distinct erosion patterns. During operation, the pure water created within the combustion chamber could be considered Hydric acid due to having a low PH and is corrosive to limestone, which also provides a clue as to the lifespan of the system's operation.

The erosion appears to show a spray or blast pattern from the Antechamber, plus a trickle down the middle where the erosion reaches a bit less than a meter in depth on the leading edge of the stone.

While the dissolved limestone would serve to buffer the PH of the water in the system, the quantity of dissolved Calcium carbonate could help to account for the Queen's chamber being found encrusted with some type of salt deposits.

The Antechamber

The King's chamber's Antechamber bears a thermal signature of having been scrubbed with hot, high velocity gasses, somewhat similar to the interior surface of an engine's exhaust manifold.

Great Step circa 1910

The King's chamber

The condition of the sarcophagus is of particular interest since a close inspection of the stone box shows extensive erosion all around the top of the box, consistent with damage from a repeated force causing micro erosions occurring over a period of time. The King's chamber south wall also appears to show substantial impact damage in the area around the southern shaft.

The sarcophagus

Image courtesy of Ancient Presence

King's chamber south wall

Image courtesy of
Ancient Presence

Chapter 11
Summary

It is this appearance of wear and tear evidence together with these aspects of physics that makes sense of the otherwise bizarre presence and physical relationships of the building's features and gives rise to the new physics based hypothesis and theory of operation for the Great Pyramid.

While Occam's razor might suggest that the simplest answer is probably true, the conventional narrative of Dynastic Egyptian tomb for the Great Pyramid is contraindicated. We know that Dynastic Egyptian tombs were typically concealed to hide their location from potential grave robbers. We also know that royal tombs featured elaborate paintings and carvings that tell about the life and accomplishments of the person entombed there.

The traditional narrative of the past was born of people from centuries ago with little to no knowledge about the concepts of physics that we take for granted today.

Viewed through the eyes of modern knowledge, the Great Pyramid presents itself as a machine. A remnant of high technology from an advanced civilization, lost to history a long time ago. While the Great Pyramid presents a solid state solution with no moving parts, the configuration and shape are proprietary since the operating strategy is dependent on gravity.

One has to wonder if we can bring this old world tech into the new world and what it might look like using modern materials, appliances and some creative configuration. Considering the recent interest in using Hydrogen as a fuel, or moving to a Hydrogen economy like the Empire of Japan, perhaps an adaptation of these technologies to produce HHO fuel, large electrical currents and high negative voltages could present useful power supply solutions in the advancement of our global energy economy.

Inscription found at the entrance of the Great Pyramid

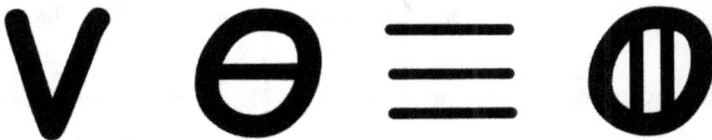

V θ ☰ ⦶

Masuline force granting feminine

Terry Lee

Chapter 12

Conclusion

The intent of this text is to advance the hypothesis as close as practical to a bona fide theory of operation, using our current level of experience and knowledge about physics. A practical set of values and estimate for the amount of heat, water and steam necessary for the system's dynamic operation is presented here.

Some of the necessary values like operating voltage levels, the efficiency of plasma HHO production, gas capacitance of the building and the practical result of electron amplification through the gaseous dielectric medium are beyond our current sphere of experience and are at best, difficult to estimate for a conclusive theory of operation.

Value set for water

One liter of water requires 2.3 kilowatts of heat to drive its phase change to produce 1.65 cubic meters of steam. One cubic meter of water will produce 1,650 cubic meters of steam.

One liter of water requires 13.2 megajoules, or 3.67 kilowatts of electrical energy for one hour to drive the electrolysis and produce 1.2 cubic meters of Hydrogen gas and 0.6 cubic meters of Oxygen. One cubic meter of water will produce 1,200 cubic meters of Hydrogen and 600 cubic meters of Oxygen.

A reasonable estimate for the system's operation can be established by plugging in the energy values and basic properties for one liter of water. The baseline presented here suggests that the system supplied sufficient heat to boil 4 liters of water per second, producing 6.6 cubic meters of steam per second to drive the electrostatic generator in a dynamic fashion. Enough steam to produce a 5 meters per second velocity of steam through the horizontal passage and equal the internal volume of the building in about 5 minutes.

Impression of operation

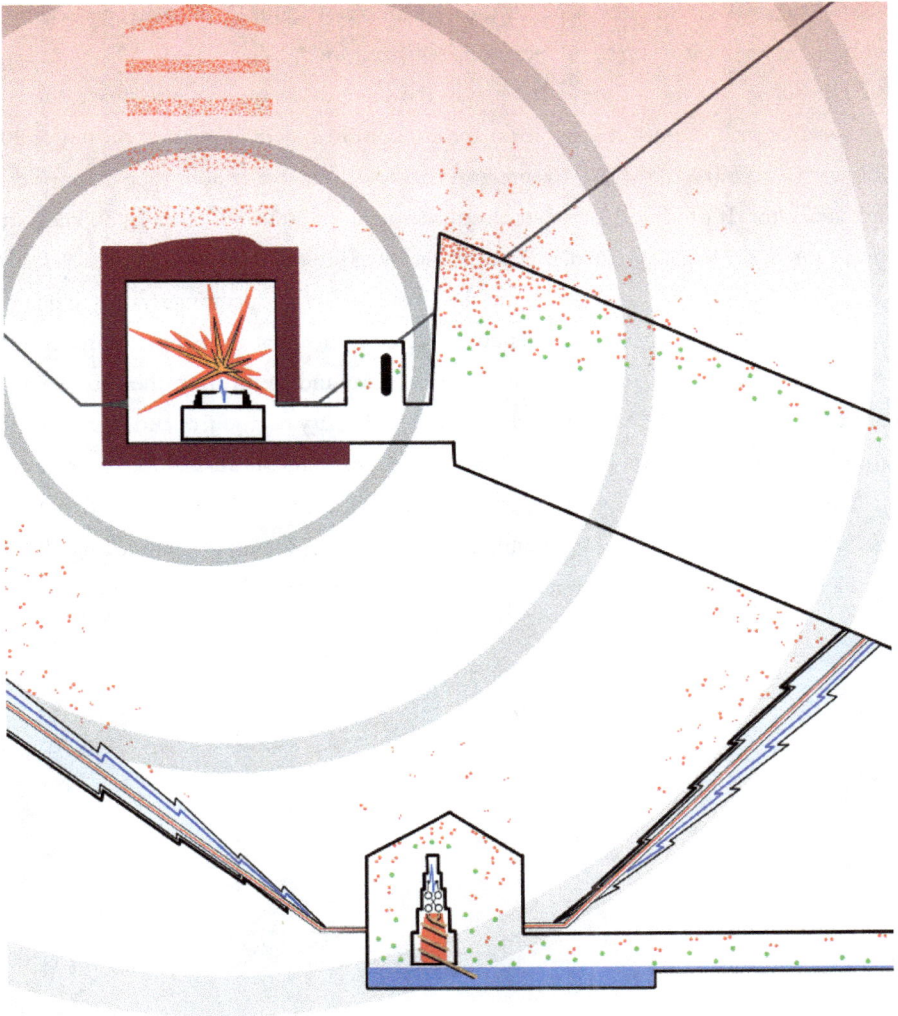

Terry E. Lee

Values for heat

The quantity of heat required to power this example of boiling 4 liters of water per second suggests a minimum value for the heat source of about 10 kilowatts per second, once the system warms up to full operating temperature.

The SNAP 10a reactor presents a handy and practical example of a compact, robust heat source with an output suitable for the application and an initial heat output of 30 kilowatts. Of course the SNAP 10a presents an early example of radioisotope thermoelectric generator. It produced a modest electrical output of 500 watts from its 30 kilowatt heat source, yielding an abysmal thermal efficiency of less than 2%.

In contrast to our current paradigm, the Great Pyramid's operating strategy suggests an advanced level of thermal efficiency by recycling the heat of its operations as hot water from condensed steam. This includes the heat and steam from the combustion chamber, along with the heat and steam from the primary heat source and operation of the electrostatic generator, all returns as hot water back to the steam chamber water pool.

Lastly, the system leverages the energy of vibration and high voltage ionization to drive electrical arcs through the steam chamber directly over the water pool, producing HHO fuel and further recycling the heat back into the water contained within the pool.

Steam powered electrostatic generator

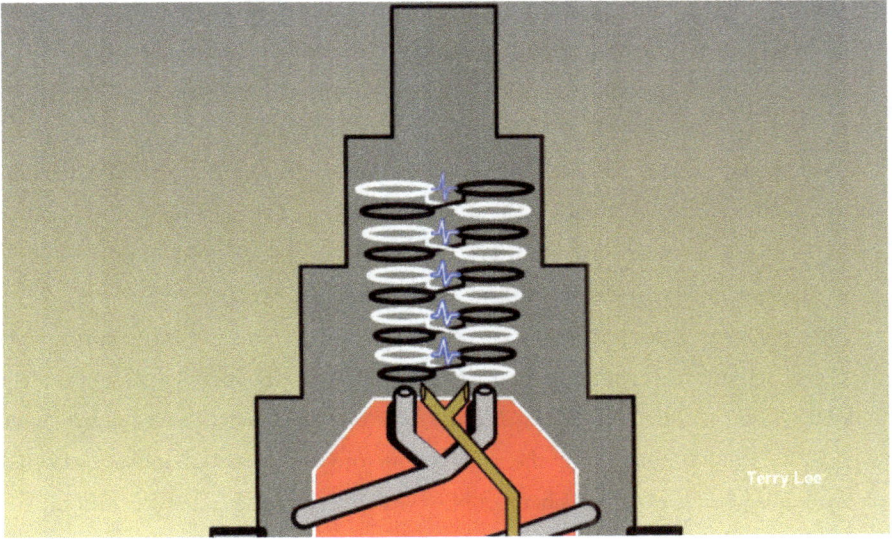

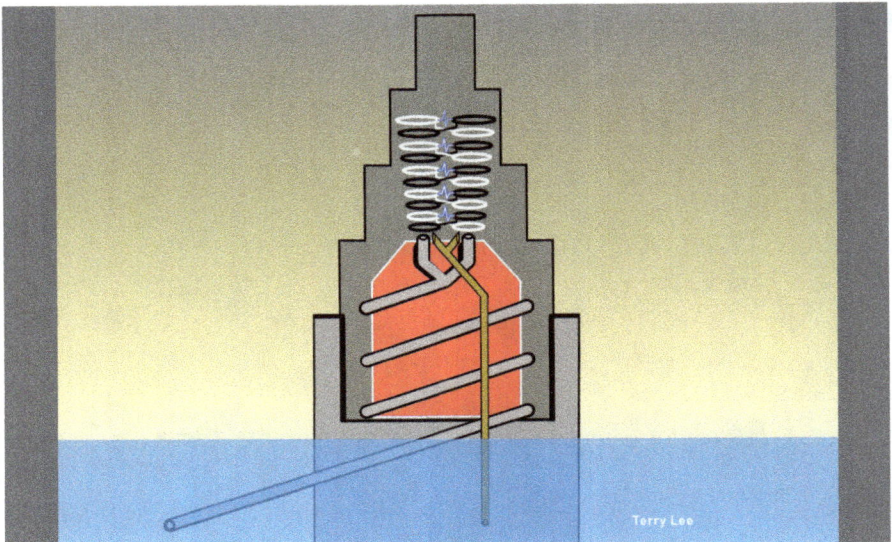

Volumes

With the approximate internal volume of the building at roughly 2,000 cubic meters of space above the level of the drain and the gas capacitance of the stonework estimated to be about the same as the proper internal volume of the building, gives an estimated total of about 4,000 cubic meters of space to be filled with steam, water vapor and Hydrogen gas.

Given the 16 cubic meters of water available in the steam chamber at system start up and the volume conversion rates for liquid water to steam and liquid water to HHO gas. It seems pretty easy to see that around 8 cubic meters, or about half of the steam chamber's water supply would be needed to initialize the building and enable system start up.

Once the system starts up and begins running, each combustion event in the combustion chamber would send a pressure wave that travels the length of the Grand Gallery and reflects off the surface of the water in the reservoir. The impact of the pressure wave would cause water to splash from the reservoir and into the Queen's chamber horizontal passage, thus keeping the steam chamber filled with water during system operation.

System components

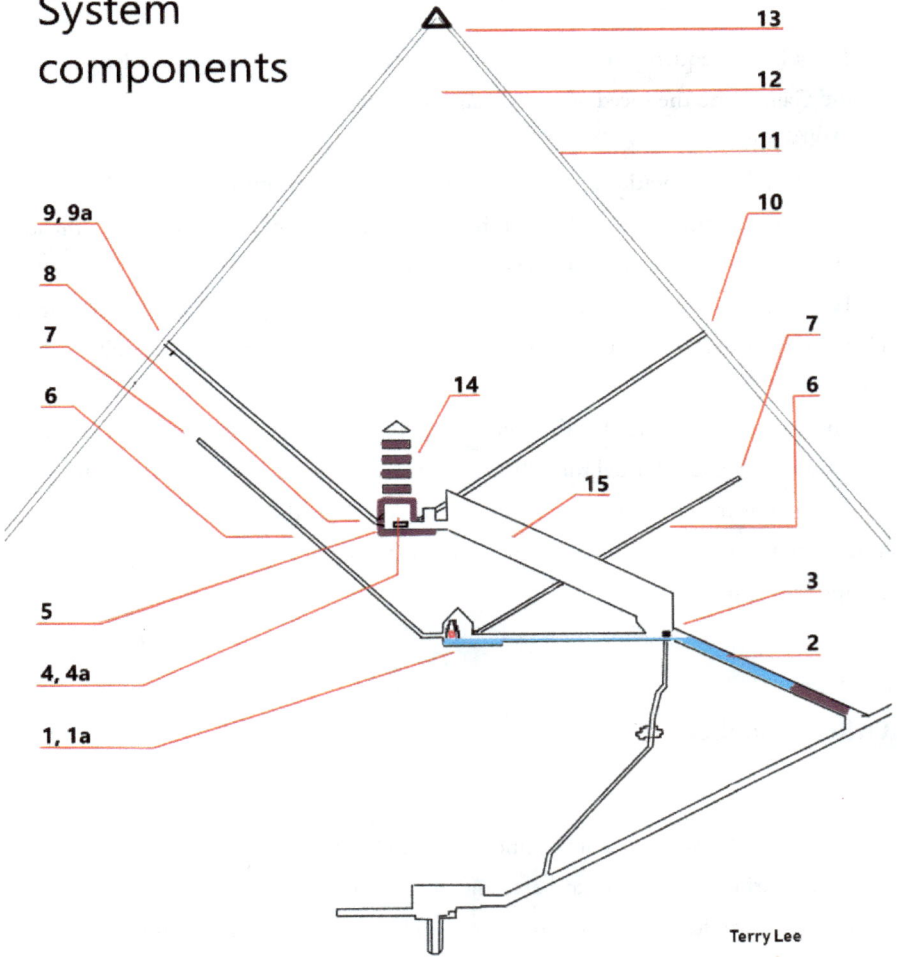

Terry Lee

Frequency values

The 4 hertz frequency of the internal combustion cycle is given by the length of the Grand Gallery and the speed of sound, adjusted for temperature and humidity to 100*C and 100% humidity.

The 3 kilohertz vibration frequency for the walls of the combustion chamber is given by the estimated thickness of the chamber walls and the velocity of the P wave through granite, which is roughly 6 kilometers per second.

The frequency of intermittent high voltage electrical arcs produced across the steam chamber is locked to the vibration frequency of the combustion chamber walls at about 3 kilohertz.

The acoustic frequency for the combustion chamber is given by its dimensions and the speed of sound, adjusted for 100*C and 100% humidity. It's curious that the east – west length of the chamber is exactly twice the width of the north – south dimension and suggests an acoustic frequency of 72 hertz north and south, while the east – west component runs at 36 hertz.

Other thoughts

The steam chamber presents a difficult and highly corrosive environment due to the constant swirl of steam, electrical arcs and elemental Oxygen. The harsh nature of the environment would call for resistant materials like Gold, or platinum group metals such as Palladium or Iridium for their high temperature and chemical resistant properties. Coil sets fashioned from an alloy of Platinum and Palladium would seem optimal for their ductility and properties as a catalyst.

Operation of the steam powered electrostatic generator would also require some means of driving a positive pressure gradient between the water intake and the high pressure steam output nozzles. Some type of solid state configuration by routing a portion of the superheated steam from the output back to the water intake.

Today's researchers stand on the shoulders of the giants who came before us and continue to be inspired by Nikola Tesla's research and pioneering work into the advanced concepts of alternating current and frequency based technologies. His truth and light led the way for development of these technologies, despite the resistance and vilification from the established direct current infrastructure of Thomas Edison.

Thomas Edison went to great lengths to prove that A/C current is very powerful and deadly dangerous. Of course he was only too right and perhaps the same is true for HHO fuel. HHO is indeed a dangerous, explosive gas with a reaction speed of 3,000 meters per second and the highest specific impulse of all known fuels.

By the same token, today's researchers into alternative, advanced or immanent technologies can also get a taste of what Nikola Tesla must have felt. The joy and immense satisfaction of discovery and technological advancement, contrasted by the weight of existing narratives and the established infrastructure.

Never the less, technology marches on and it seems only a matter of time until water finds its place as a fuel stock of choice for clean, sustainable fuel once our technology catches up and finds its niche to supply our energy needs for tomorrow.

With this promise of the future in mind, I'll close with the following thought. The only equation suggested in this book is *"Townsend's equation for current growth in a gaseous dielectric medium"* and the only formula presented is the formula for water. H^2 plus O equals water and the highest specific impulse of all known fuels.

Terry E Lee

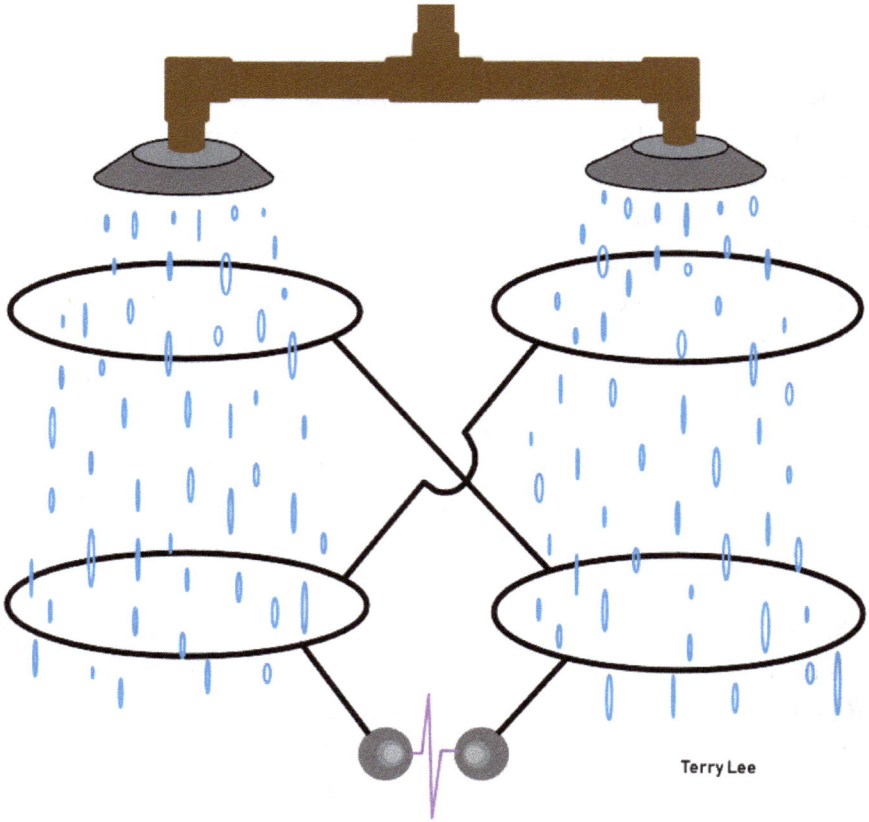

Terry Lee

Kelvin's Water Dropper experiment

Also known as Kelvin's Rainstorm, or Lord Kelvin's Thunderstorm Generator. The experiment shows that dribbling water droplets through a set of cross connected coils will produce high voltage electrical arcs of 10kv – 20kv between the electrodes.

Piezoelectric current

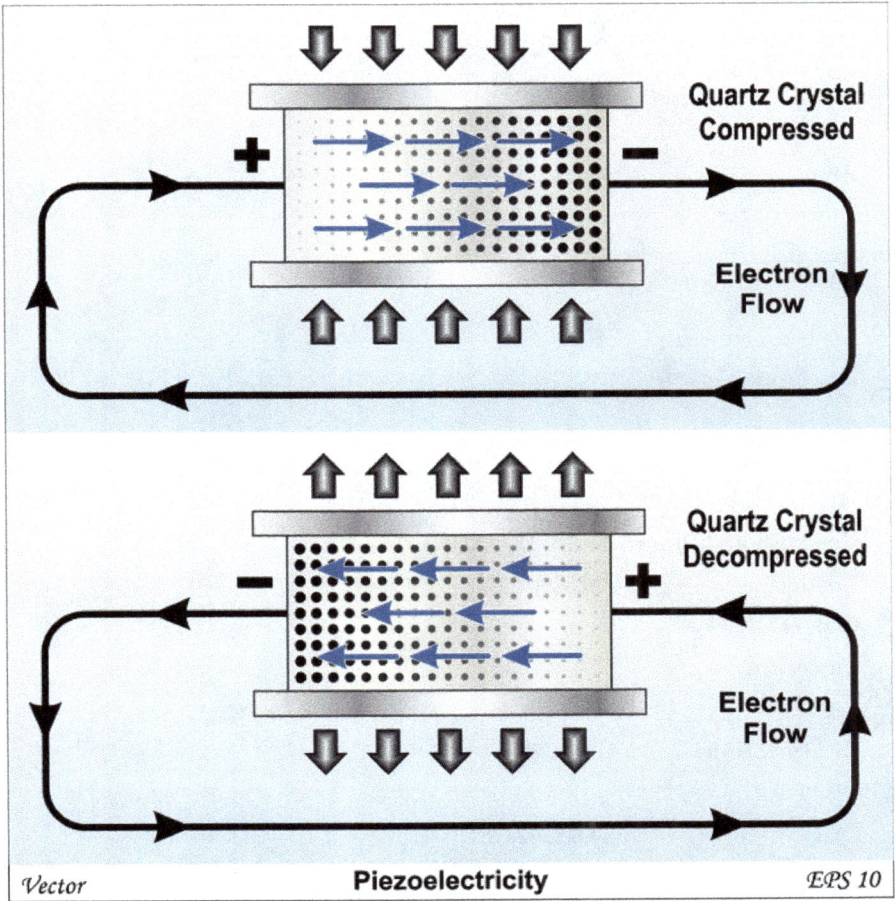

Quartz Crystal
Compressed

Electron
Flow

Quartz Crystal
Decompressed

Electron
Flow

Vector   **Piezoelectricity**   *EPS 10*

Print index

Print index

Print index

Print index

Print index

Print index

Chapter 13
Bonus pics

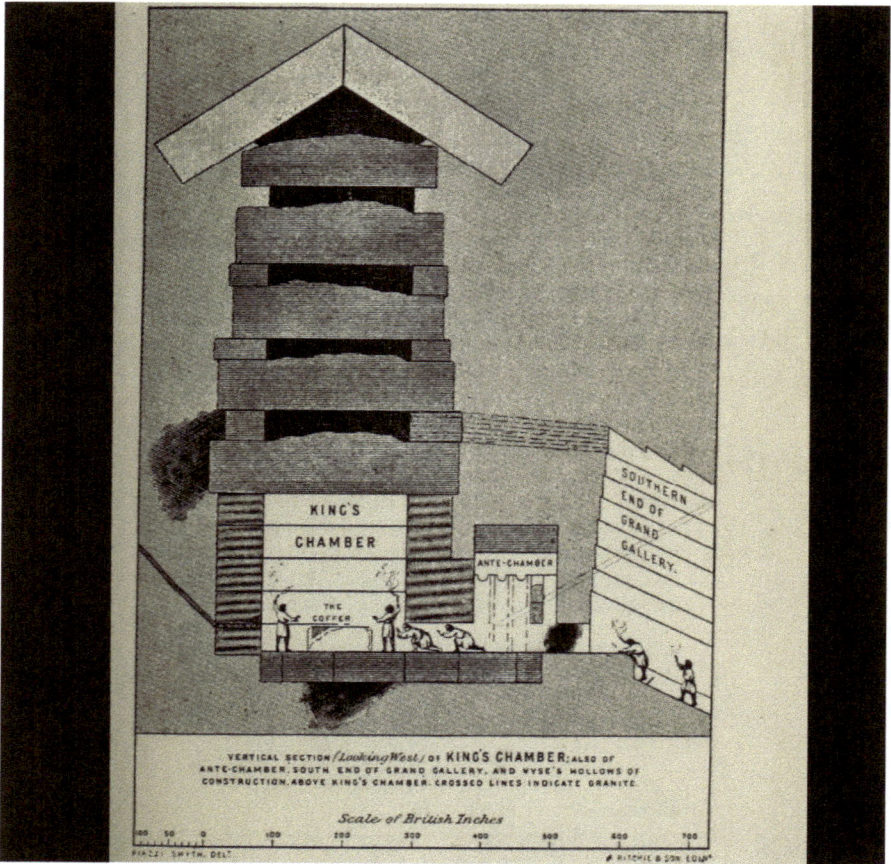

VERTICAL SECTION (*Looking West*) of KING'S CHAMBER; ALSO OF ANTE-CHAMBER, SOUTH END OF GRAND GALLERY, AND WYSE'S HOLLOWS OF CONSTRUCTION, ABOVE KING'S CHAMBER. CROSSED LINES INDICATE GRANITE.

Scale of British Inches

Broken drain circa 1900

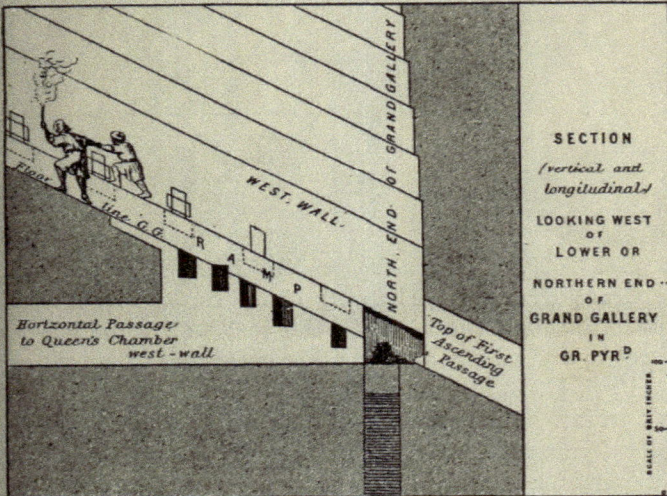

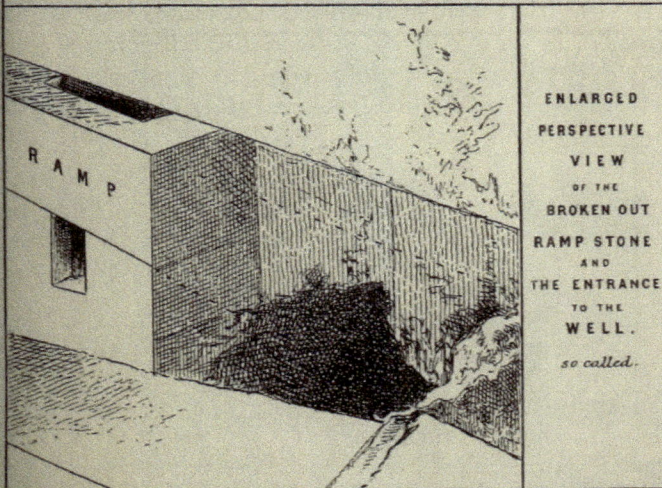

Tesla's vision

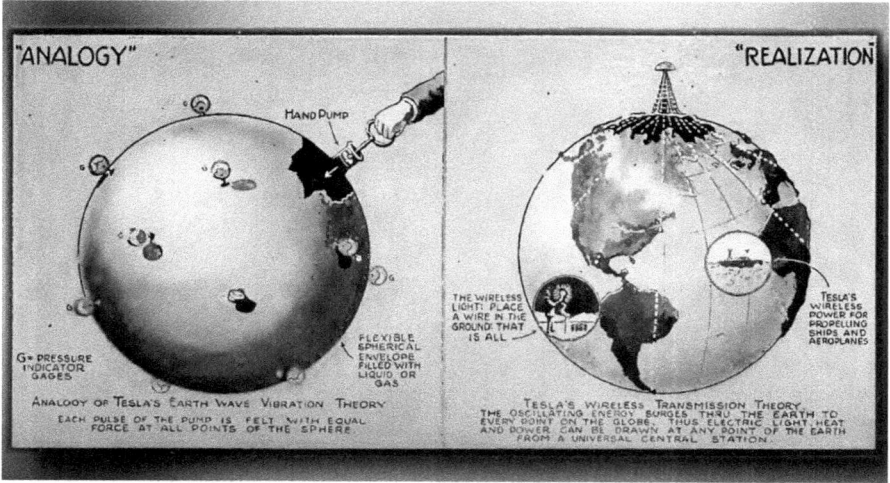

Left panel

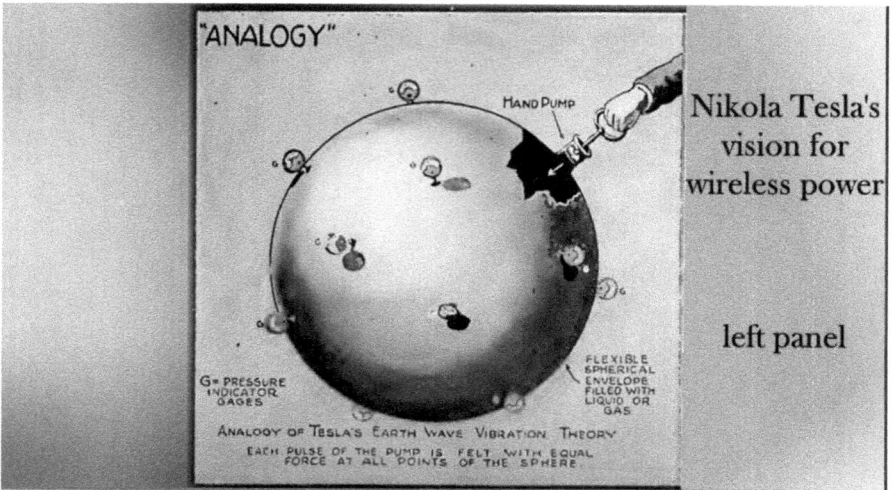

Wardenclyffe tower

Right panel

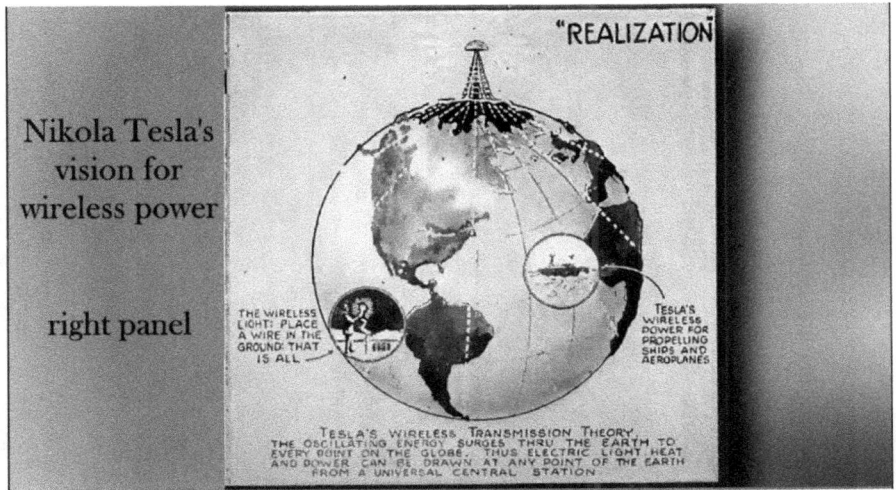

Nikola Tesla's vision for wireless power

right panel

"REALIZATION"

THE WIRELESS LIGHT: PLACE A WIRE IN THE GROUND: THAT IS ALL

TESLA'S WIRELESS POWER FOR PROPELLING SHIPS AND AEROPLANES

TESLA'S WIRELESS TRANSMISSION THEORY: THE OSCILLATING ENERGY SURGES THRU THE EARTH TO EVERY POINT ON THE GLOBE. THUS ELECTRIC LIGHT HEAT AND POWER CAN BE DRAWN AT ANY POINT OF THE EARTH FROM A UNIVERSAL CENTRAL STATION

Karnak, Egypt

Axum, Ethiopia

Osiris device at Abydos

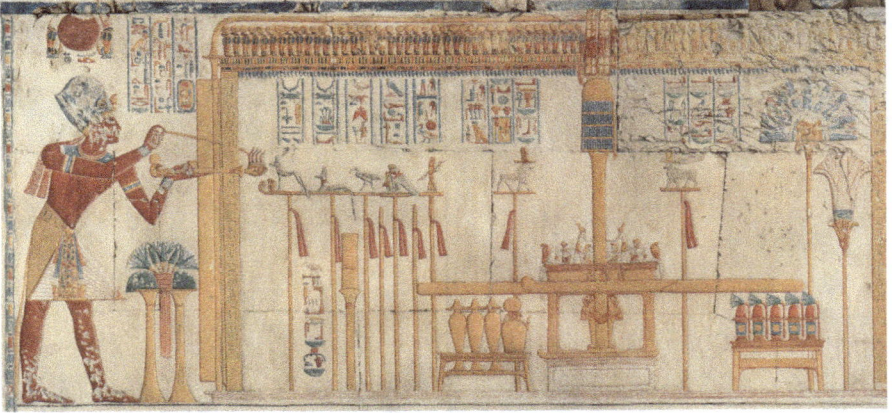

High voltage fishing in ancient Egypt

Cusco, Peru

Sacsayhuaman near Cusco, Peru

Hilltop at Sacsayhuaman

Suspected resonant cavity device

Notes:

Notes:

Coming for 2026

Ancient Egyptian Technologies

Mark Cronin

Advanced research into the energy infrastructure of ancient Egypt

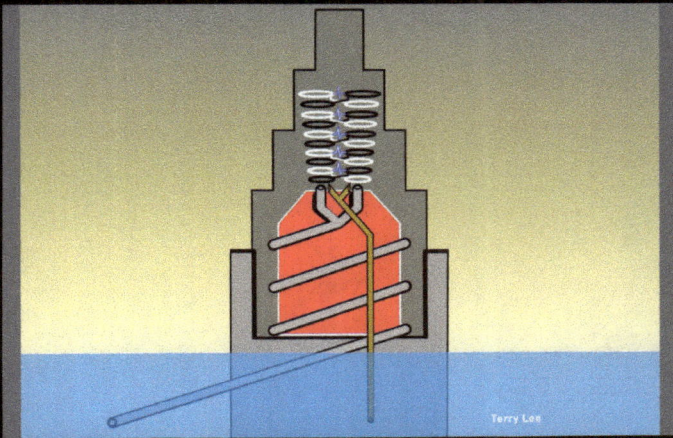

About the author

Come along for a deep dive into the Hydrogen technology
of our distant past with a detailed description of the Great
Pyramid's strategy for producing incredible amounts of wireless
power from water. Presented by a career automotive tech and long
time student of the building, Terry Lee. A former ASE certified master
auto technician with an L1 certification and an interest in things electrical